# ENERGY SHOCK

Also by LAWRENCE SOLOMON
*The Conserver Solution*

From POLLUTION PROBE/ENERGY PROBE
*Additive Alert* (by Linda Pim)

# ENERGY SHOCK

## After the Oil Runs Out

*Lawrence Solomon*

An Energy Probe Project

1980

DOUBLEDAY CANADA LIMITED
*Toronto, Ontario*

DOUBLEDAY & COMPANY, INC.
*Garden City, New York*

*Library of Congress Catalog Card Number 80-1072*
*ISBN 0-385-17160-9 (hardcover)*
*0-385-17161-7 (paperback)*



*Printed and Bound in Canada by The Webcom Company Limited*

*The paper in this book contains post-consumer waste.*

---

*To Delicia*

# FOREWORD

I have always been intrigued by the ability of politicans and other actors in the public and private spheres to concentrate on the small and irrelevant. Ask us to worry about a stop sign and we'll gladly comply. But ask us to review our policies about energy use, or its ownership or distribution, and we'll be evasive. It's natural, of course, to tackle the problems that we can't solve alone in a political time-frame.

This book tackles the large and significant issues, and, as some have suspected all along, concludes that apparently unmanageable problems can be solved if we return to basics.

Human beings have had great success integrating their individual needs with their neighborhoods and cities. Here are the breeding grounds for popular culture, industrial innovations, the realm of serious thought, and political movements. Yet in recent times we've turned our backs on neighborhoods and have chosen to vest increasing authority in large-scale government and remote bureaucracies. In doing so, we have lost grip on many institutions and now find it difficult shaping them to meet society's changing needs. The inflexible economic and social structure that has resulted is leading to a stagnant economy and insecure society.

As this book points out, it's that inflexibility that needs changing. Those of us who worry about governing cities will have to give special care and attention to scales of efficiency in urban areas, a field we have shunned too long. We'll have to find new ways to involve residents in decision-making. We'll have to restore the ability of cities to control those areas that concern them most. We'll have to make cities less dependent on outside food and fuels.

This book asks us to let the sun have its proper place in the city, as sustainer, as democratizer. It's strange that we have to rethink the role we've assigned to our friend the sun, but we haven't much choice. Apart from solar energy, the only things the city does not import are human talent and innovation. Using these two products, and books like this one, perhaps we can make self-sufficient cities that we're proud to live in.

*John Sewell*
*Mayor of Toronto*

# ABOUT ENERGY PROBE

One of the most effective environmental organizations on the continent, Energy Probe first made headlines coast to coast in 1975 with its exposé of radiation contamination in the town of Port Hope, Ontario.

The national scandal forced the federal government to clean up the hazard and Energy Probe has had little trouble getting its message heard ever since. Its books are profoundly influential pioneers in their fields; its economic studies are published by bodies like the Economic Council of Canada, and their recommendations are endorsed by leading economists like Richard Lipsey and Douglas Hartle. Energy Probe's message—through the newspapers, radio and TV—is heard an average of one-and-a-half times a day.

That message has an uncanny record of being right.

In 1975, Energy Probe pitted its planning expertise against that of Ontario Hydro and correctly predicted that the giant utility was engaged in a massive overbuilding program—a realization that the government took five years, and $10 billion, to discover. (By 1979, official bodies like the Royal Commission on Electric Power Planning and the Select Committee on Ontario Hydro Affairs were preferring Probe's projections to those of Hydro.)

When government and industry scoffed at Energy Probe's assertion that conservation could make huge inroads into Canada's energy consumption, it decided to make believers out of both by taking an old, energy-inefficient house in downtown Toronto and economically eliminating 85 percent of its energy requirements. (Now called Ecology House, it is open to the public.)

Through meetings with political leaders and heads of crown corporations, through its public-speaking activities across Canada and abroad, and through approximately 50,000 responses a year to requests for information, Energy Probe has acquired a grudging respect from its adversaries and a selfless dedication from its many volunteers and financial supporters. Those wishing to advance this public interest group's work are welcome to call or write.

Lawrence Solomon is quite approachable, and can be reached at (416) 978-7014:

Energy Probe,
43 Queen's Park Cres. East
Toronto, Ontario M5S 2C3

Energy Probe is a project of the Pollution Probe Foundation. Donations to Energy Probe are tax-deductible.

# CONTENTS

# INTRODUCTION
## *Nuclear Earth*

WERE THE SUN to burn out, human life on earth would not necessarily end. The earth, its center a molten mass, gives off a great deal of heat, but only enough to keep our homes at a bare -240°C. The coal, oil, and natural gas used to raise the temperatures of our homes only a few degrees from 0°C to 20°C on a winter day would now be needed by those who had survived to raise home temperatures by 260 degrees every day of the year, for as long as these depleting fossil fuels lasted. The anti-nuclear lobby defused, existing nuclear plants would provide enough heat and light for limited greenhouse food production as well as for home needs. Should we develop advanced nuclear breeder systems and fusion reactors, humankind might once more multiply and prosper.

Lifestyles would be different, however. The winds, the rivers, the rains and storms—phenomena all powered by the sun—would cease. Night would no longer turn into day, summer would no longer follow spring, the cycles around which our lives are presently structured would be flattened out—perfectly horizontal. There would be little sense in going to bed at midnight if the sun didn't come up at 6 A.M.; little sense in having 365 twenty-four hour days when our years could be measured much more understandably in some decimal system. Eight-hour work days would be replaced with work periods, scheduled continuously, to fully utilize our machinery and other capital equipment. No work period would be more desirable than another due to antiquated habits of working from dawn to dusk.

Earth, which had been powered by the solar engine that grew our grains and recycled our rainwater, would now be powered by nuclear machines that were not tied to traditions like autumn harvests, that could dispense with nature's cyclical and cumbersome way of providing pure $H_2O$. Nuclear earth would find a way to distribute water using contrivances other than soil and clouds. Nuclear earth would select harvest dates that best met its own production schedule. As profoundly as man was forced to conform to the rules of the solar society, so too would man need to obey the new structure that had become his life source.

Transportation and outdoor work would be immensely difficult in a permanently frozen environment. Human colonies would need to build nuclear reactors on top of the uranium fields that fueled them, and human civilization would need to huddle around its nuclear plants. Because heat can't be efficiently transmitted great distances, nuclear complexes would tend to be spherical, with less heat available at the outer edges, where the less privileged might live. Because temperatures underground would be higher than temperatures at the surface, we would become a burrowing people, living in subterranean luxury to bask in the glow of the earth's core. Taking advantage of the laws of physics, the high-rises above ground would be designed to trap channeled heat rising from below.

Though we would be too sophisticated to worship the reactor as primitive tribes worshipped the sun, the reactor, as the source of all heat and light, would nevertheless become a sacred institution. An accident, sabotage, a simple malfunction could spell the end of the colony. Society could afford to take no chances where nuclear power was concerned: only the most talented and disciplined would be chosen to work in the plant; only the most privileged and trusted would have access to it. Those controlling the most information about nuclear power would become the most important people, their knowledge literally the difference between life and death for all. Were an aristocracy to emerge, it would be a nuclear aristocracy whose pronouncements on scientific and social matters relating to nuclear technology would be accepted as gospel. The opinions of those less schooled in nuclear thought could not be given much importance in a vulnerable society preoccupied with survival; and wisely so: we couldn't afford much dissident opinion or any social unrest. Humanity's history by then would be replete with tales of colonies that had failed to secure their existence: the loss of our freedom of speech, or other individual liberties, would be welcomed when opposed by the larger consequence.

Society would become, in effect, highly centralized, with the central nuclear power system the basis of our social structure, economic structure, and political structure. Few variations in lifestyle would evolve, even among disparate nuclear communities: the civilizations that survived in Tokyo or Cairo would be little different from those in Montreal or Los Angeles because culture would no longer be based on regional adaptations to variances in climate or geography. Culture would be based on universal adaptation to the uniform properties of the atom.

Doomsayers aside, current estimates have our solar system lasting for several billion years. Our sun is not about to burn out. Yet in many ways we're acting as if it already has. Rather than base our economies on solar fuels that are part of nature's cycles, this century has seen increasing use of non-renewable fuels that are independent of cyclical constraints. Agriculture has seen large infusions of petrochemical fertilizers; water and wind energy no longer form the basis of our transportation systems. Farmers across the continent abandoned their windmills and everyone's dependence on locally available wood fuels declined. By degrees society has become slowly centralized: more and more dependent on electrical grids vulnerable to blackouts, more and more dependent on long supply lines for what has come to be considered our primary fuel—petroleum. The cost of abandoning the security of nature's economy for one based on non-renewable fuels—and so one destined to self destruct—is now being felt in Windsor and Detroit with the collapse of our fossilized automobile industry. Soon it will be felt in all metropolitan areas as fuel-short suburbanites flee back into the city.

The phasing out of conventional fossil fuels is inevitably leading to profound structural changes in the economy, to profound structural changes in society. The shocks are reverberating throughout our human systems, affecting the way we eat, clothe ourselves, conduct trade with our friends, conduct war with our enemies. The shocks are exacerbated by the strains of trying to accommodate the two entirely incompatible energy forms vying to replace petroleum's place of pre-eminence. One form fits in perfectly with nature's cycles, adding no new energy burden for the system to bear; it harnesses only freely available energy by tapping the efficiencies in the system. That energy is conservation.

Conservation, in refusing to waste the immensely abundant energy in the sun, wind, and other sources, does no more than redirect energy to useful purposes. Conservation energy is the short form, the catch-all phrase, for the sum total of tapped solar, wind, water, and other renewable forms of energy as well as the registered efficiency gains in the use of non-renewable fuels like oil (which takes eons to be formed) and nuclear power (which, unless it can be stretched out by advanced systems, is only expected to last thirty to fifty years, or as long as uranium supplies last). Because conservation is available in infinitely diverse forms everywhere in the world, conservation is a component of human diversity, culture, and individuality. For the same reason of infinite diversity and thorough decentralization, conservation is difficult to monopolize, becoming a safeguard for individual freedoms.

Opposing conservation is nuclear power, an energy form with immense potential as an electricity provider but no prospect of acting in concert with other systems under the sun. Nuclear power is an add-on source, unleashing energy the earth had kept under control and creating as waste products potent energy forms like plutonium previously unknown to nature. As highly centralized as conservation is decentralized, nuclear power is too demanding to exist in conjunction with its competition for long, despite nature's craving for diversity. Not a self-sustaining complement but a highly dependent aberrant in earth's biosphere, nuclear power can only be seen as a mutant. Nuclear may be allowed to live out its remaining years, but it can't survive and reproduce except to the ultimate exclusion of diverse energy use.

The precise qualities which make nuclear power highly centralized make it highly monopolizable. The energy form society adopts will direct our future way of life, but powerful, highly centralized corporations are poised and ready to determine the energy form for society. Corporate decisions could irreversibly predetermine the course of mankind. In an ideal free market, where corporations had to obey the law of supply and demand, the influence of the corporation would not be a concern—the economy would register the dollar votes of the public with unerring accuracy, and the decision would be made democratically.

But we are far from having a free market: powerful public and private monopolies control oil, electricity, and other major energy sources, interfering in the energy marketplace where conservation and renewable forms of energy are trying to compete. We are heading toward two simultaneous energy futures; one being promoted by nature's efficiencies, the other by business exigencies, but both governed by common metaphysical laws. In the same way that members of society strive for answers to the questions society asks of them, members of corporations strive for answers to corporate dilemmas. If what was good for Exxon and Ontario Hydro were good for their countries, there would be no conflict—the questions posed and the answers found by each monopoly would approximate those of their societies. But corporate citizens and human citizens persist in operating from different premises: humans still work to cycles—still plan on their next summer's vacation, still work toward their next generation's welfare—while corporations work toward eliminating nature's cycles, favoring round-the-clock production runs and quick returns on investment. Corporations, by their nature, run counter to the interests of individuals. Their value to humans comes not from their methods of operation but from the material product of their operation. The arrangement generally is satisfactory as long as the methods do not influence the products.

But this cannot be helped in the case of corporations involved in energy, because energy use, and the technologies dependent on energy use, cannot be separated from the way in which we live.

Corporations have once before usurped society's right to determine its own lifestyles. This happened around the turn of the century, when our patterns for using fossil fuels and electricity were determined. Because corporate production of fossil fuels and electricity is now all but saturated using conventional means, new means are being sought. These means are highly centralized; they are in conflict with means that are highly decentralized; and they are in conflict with nature's cycles, and so face nature's formidable resistance.

Mankind is now at a turning point, potentially the last turning point because the nuclear and solar options cannot co-exist; each will seek to squeeze out the other, and each has the potential to indefinitely provide 100 percent of our energy requirements.

We are being asked as a society to vote for the form of energy we'll turn to after the oil runs out. Here we'll determine whether the traditional family is strengthened or weakened; whether the city is to be saved or scrapped; whether we are to drift toward a freer or more totalitarian society; whether any of our social, economic, and political institutions are to survive, and in what form. Yet, though voting day is approaching, very few people have been handed the ballots that will give them a say in how they and their descendants will live. The public is responding as other disenfranchised constituencies have—by forming alliances and demanding their rights.

The battle lines are now drawn, the conflict defined: it has come down to nuclear vs. nature. The immovable centralized object is confronting the irresistible decentralized force, and we are all feeling the energy shocks of their impending collision.

# PART 1

# Power Brokers

# 1

## *World in Transition*

SOLAR ENERGY, ALTERNATIVE education, housing co-ops and no-name foods are not unrelated phenomena but part of one of the great transitions in human history—the shift away from non-renewable forms of energy. Like the previous transition into coal (which brought us the Industrial Revolution) and oil (which brought us automobiles and plastics), the new energy technologies emerging to replace those based on our dwindling supplies of fossil fuels will totally dominate our lifestyles. Society's mobilization in response to the energy crisis has led to frantic development of new energy forms. One of them is in convincing lead. That energy form is conservation—so quick off the starting mark that it has already led to the cancellation of over 200 nuclear reactors in Canada and over 2000 in the United States scheduled to be built over the next twenty years.

Before the OPEC oil embargo of 1973, there were no serious attempts at energy conservation. Within five years conservation was the fastest growing source of energy in North America, recording staggering savings in energy in every sector of society. New office buildings, like the Ontario Hydro headquarters in Toronto, had eliminated the need for 65 percent of normal energy requirements and its office space cost no more than space in less efficient buildings. Factories, like the Bata Shoe Company in Picton, Ontario, had cut out 90 percent of their oil bill by redesigning existing plants. So successful was Bata's experience that it began looking at its other ninety-seven plants around the world with an eye to energy conservation. Private homes, like those being built in the extreme cold of Saskatchewan, had eliminated up to 90 percent of energy requirements. Constructed at a competitive price, these houses proved so popular developers were unable to build them fast enough.

Conservation continues to lengthen its lead over nuclear power. Within ten years of the OPEC oil crisis conservation could force the nuclear industry to cancel a total of 5000 reactors that they were planning to build over the next fifty years. Within thirty years conservation and solar power could replace our entire dependence on the energy we now get from nuclear, coal, oil, and natural gas.

Such a rapid transition to a different form of energy is not unprecedented. In fact, it would follow the pattern of history.

There were no oil wells before 1859. Within five years of the first strike at Titusville, Pennsylvania, dozens of oil refineries had sprung up. Within ten years oil was a common commodity in homes, offices, factories, and farms. Within thirty years oil was being delivered to every corner of the world—Sumatra, Siam, Siberia, the Sahara—on the backs of native bearers, on camels, by oxen, by sampans.

There were no electric generation stations before 1882. Within five years of Manhattan's illumination dozens of generating stations had sprung up. Within ten years electricity was a common commodity, serving hundreds of electric railway companies, thousands of industries and millions of individual consumers. Within thirty years the continent was successfully electrified, benefitting virtually every city, town and village.

With the same speed, new technologies based on the new energy forms received routine acceptance from a continent unfettered by Old World prejudices.

Two years after Morse sent his first electric telegraph message from Washington to Baltimore in 1844, the Toronto, Hamilton, Niagara, and St. Catharines Electro-Magnetic Telegraph Company was transmitting messages between those Ontario points. A few months later, lines were extended to Buffalo to link the burgeoning U.S. and Canadian grids. One more year and the Montreal Telegraph Company had strung its wires from Toronto to Quebec, and the British North American Electric Association was in the process of linking Quebec to the Atlantic Coast.

The first electric streetcar systems in North America began service in 1887 at St. Catharines, Ontario, and Richmond, Virginia. Within three years the technology had reached all the way to the West Coast, with Victoria and Vancouver installing electric lines. Three more years and General Electric emerged as a major manufacturer of electric locomotives. By the turn of the century almost every city on the continent had its own electric streetcar system.

The automobile also won speedy acceptance. Around 1900 only one person in 10,000 owned one of the new "horseless carriages." By 1930 car purchases had accelerated so quickly that there was one car on the road for every five people—or just about one per family.

Solar TVs are now in commercial use in Africa, solar wrist watches are on sale in North America, and manned solar flights are surpassing the rate of development that followed the Wright Brothers' earlier success at Kitty Hawk.

The speed with which the whole of society can change is mind-boggling—too mind-boggling for our heavily bureaucratized government and corporate leaders to conceive of, or allow for, though the speed has been demonstrated time and time again. The human appetite for progress has never been successfuly restrained for long.

Computers, television sets, hundreds of household appliances, countless industrial advances—all have been introduced quickly and efficiently with generation after generation of improvements soon following. The colossal, clumsy computers of World War II vintage were refined and condensed into sleek hand-held calculators; snapshot-sized black-and-white TV sets have evolved into a flourish of different-sized color models that are used for private entertainment or public education.

Revolutions of the profoundest type occurred; lightbulbs and streetcars, motion pictures and radios, have changed our lives at breakneck speeds. Society has been transformed without public outrage at the new kinds of energy, partly because progress has always been prized, partly because events and inventions were able to adapt to our needs (and our environment's needs) too quickly and flexibly for us to be able to react.

Energy forms, and the technologies dependent on them, have all exceeded the wildest predictions of their early enthusiasts . . . with one exception. Nuclear power, despite forty years of dedication from some of the best minds in the world, has made only pathetic progress toward its goal of meeting all our energy needs. In retrospect the visions of the nuclear prophets seem to us today to be no more than the ravings of true believers.

Writing in *Popular Mechanics* in 1941, a professor at the California Institute of Technology predicted the advent of atomic automobiles. A single pound of uranium would provide the energy of 250,000 gallons of gas, he enthused, allowing cars to go 5 million miles between refuelings. His claim that car engines would become the size of typewriters was scoffed at, however, by the president of the Society of Automotive Engineers, who predicted more compact atomic auto engines the size of a person's fist.

The problem of transportation solved, everyone's favorite subject of conversation—the weather—could be tackled. Artificial suns made from chunks of uranium mounted on towers could bring the climate under control. "No baseball game would be called off on account of rain in the Era of Atomic Energy," wrote the author of *Atomic Energy in the Coming Era.* "No airplane would by-pass an airport because of fog. No city would experience a winter traffic jam because of heavy snow. Summer resorts would be able to guarantee the weather and artificial suns would make it as easy to grow corn and potatoes indoors as on the farm." "Weather modification," Edward Teller and other leading scientists wrote, could also be achieved by using "the debris from a nuclear explosion [to] seed rain clouds."

Other optimists, like the president of a chemical association, predicted nuclear batteries were going to run everything from home washing machines to wristwatch radios. Scientists predicted miniaturized nuclear power plants would become a boon to the home hobbyist. Even our bodies would benefit—atomic energy would be a means of improving our health, increasing our sexual fertility, and prolonging our lives.

None of the nuclear fantasies have come to fruition. Despite the hundreds of billions of dollars that have been poured into the nuclear industry by both governments and corporations, nuclear power is all but insignificant to us today as an energy source, meeting only about 1½ percent of our energy needs (or less than that provided by firewood). The failure over the last forty years cannot be attributed to the anti-nuclear protest movement. For the first thirty of those years nuclear power had no opposition, embraced by all as the best hope for mankind. The failure cannot be blamed on the biases and obstructionism of environmentalists. Until the early 1970s nuclear power was promoted by them as a clean alternative to coal. The failure cannot be blamed on undue caution by the nuclear industry in protecting its workers or the public from health and safety hazards. Only recently have the social obstacles posed by environmentalists and consumer advocates complicated the simple desires of the nuclear industry.

Business acumen or scientific expertise also cannot be faulted. Unlike earlier energy forms and earlier energy technologies—often stumbled upon by entrepreneurs who started from scratch to form the Fords and Exxons of today—full-blown companies like General Electric and Westinghouse were brought in on the nuclear industry from the start to join the best scientific brains in the business in beating Hitler to the first atomic bomb.

The failure of nuclear power can only be blamed on itself, on the inherent and often overwhelming obstacles confronting any overly centralized system. Despite receiving the benefit of every break, the support of every social sector, nuclear technology is too cumbersome, too inflexible, too dependent on external support systems to be part of any but the most desperate of energy strategies. Not even Wilbur and Orville could have gotten it off the ground.

Nuclear power is not alone in being too much of a dinosaur to have much staying power. Normally manageable energy forms which were developed a century ago with stunning speed became unmanageable when they grew to mammoth proportions. A single tar-sands plant can take ten years to squeeze out the first drop of oil; oil-shale is in the same ballpark. That new tar-sands and oil-shale projects are foundering, along with the Arctic gas and oil pipelines, should come as no surprise. Projects of their scope cannot avoid the complications that come with gross scale.

There are financial complications—sums needed are of a size sufficient to take Exxon's breath away. There are manpower complications—thousands of the skilled personnel needed do not exist, making it impossible to proceed with more than one Alberta tar-sands project, or one part of an Arctic pipeline, at a time. There are resource concerns—to keep a new nuclear station in fuel can require the discovery each year of a new place to put the additional 500,000 tons of wastes generated by the mining of uranium. There are environmental concerns—the water needed for the Midwest's shale projects is not available without cutting off drinking supplies for cities or irrigation supplies for farms. And there are basic concerns about the very sanity of the schemes: two metric tons of oil sands have to be moved with giant mining machines to produce a single barrel of oil; a radioactive inventory of 1,000 Hiroshima bombs is kept in a single reactor for the purpose of boiling water to turn the electric turbines.

As the resources needed to meet the needs of these huge energy projects are redirected from their traditional uses, society itself becomes redirected. Resources worth $7 billion are being funneled into the next nuclear plant at Darlington, Ontario—resources equivalent to the consolidated assets of a giant organization like Bell Canada and all its subsidiaries.

The investment for each new nuclear station displaces equal investments society could be making in manufacturing industries, or education, or housing. This leads to severe shortages in other fields as society turns more and more of its efforts into energy production. The stresses this has created have led to the rise of alternative institutions and alternative lines of thinking to try to cope with resource-short situations. Even conventional capital markets are being exhausted by energy investments, forcing utilities like Ontario Hydro to turn to supplementary sources like the Canada Pension Plan for additional loans while simultaneously asking their government owners for a fair share of the general tax revenues.

With investments in energy approaching 50 percent of all business investments, we are returning to the primitive era when most of our effort went to fulfill our basic needs; but now, instead of searching for firewood, we're digging holes in tar sands and mixing cement for nuclear mausoleums.

Then as now, energy was not valued in itself, as a car or a television set is prized: the less often a car needs to be refueled and the less energy a home consumes in winter, the happier we are. Yet instead of making much more efficient investments in energy conservation—which can often do the job at one-tenth to one-twentieth the cost—society spends its resources on energy production. That nuclear power produces 1.3 percent of the energy used in Canada does not prevent it from receiving 70 percent of the federal government's research dollars. That conservation and solar technologies are far superior to large-scale production need not prevent them from being blocked—bought out and monopolized by opposing interests.

This buy-out is now proceeding, but not out of a nefarious desire to see the destruction of society. The corporations involved are only pursuing the policy that best maximizes their profits and best perpetuates themselves. The buy-out and monopolization are also nothing new. The very same corporations have faced up to earlier challenges in the very same way, using the very same techniques, and for very good reasons. Large, centralized corporations and the bureaucracies they require do not easily change: they are steam rollers that can stop and start with great flair but can only deal with obstacles in their path in unyielding fashion.

This bureaucratic imperative explains the perfect predictability of large concerns: why their future methods of operation must mirror their past; how it is possible for the nuclear industry to have learned so few lessons over the last forty years; why the president of Atomic Energy of Canada still runs around trying to talk us into installing mini-nuclear reactors in shopping centers and housing complexes while Ontario Hydro's public relations department keeps cranking out flyers implying radiation is good for you.

The centralized nature of bureaucracies also explains why we are not having a smooth and rapid transition into fuels alternative to petroleum, why the normal phase-in period for new energy technologies is threatened, why massive and inefficient reallocations of funds away from social services to the resource sector are underway, and why people from all walks of life are deciding they're mad as hell and not going to take it any more.

The best that can be expected of a decision-making team (or an individual) with wide responsibility is one or two good solutions to a problem. When a problem requires hundreds, perhaps thousands of different solutions, it becomes a physical impossibility for one team to devise and apply diverse solutions appropriate to each case. Rather than passing down power to the level where the problems rest, the team's tendency is to apply the same solution to all cases. The resulting friction creates a lot of heat.

Prime Minister Trudeau's well-meaning plan for a bilingual Canada—to turn every English-speaking Canadian into a French-speaking Canadian and every French-speaking Canadian into an English-speaking Canadian—was predictably opposed in all those areas where Canadians would be required to learn a skill they'd never have occasion to use. Busing in the U.S., though intended to redress a serious inequity, only redistributed social injustice by seeking a blanket solution to a multi-faceted problem.

By definition, a centralized system involves some form of social or economic subsidy: people in one place must accommodate themselves to the needs of people in another place, presumably in the long-term interests of all. A centralized education system, where the same curriculum is taught to millions of children, requires parents to give up individual preferences to achieve a uniform standard. A centralized economic system, where resource-rich areas are required to buy back in manufactured form the raw materials they earlier had shipped to industrial areas, requires the people of one region to give up benefits in the hope of a higher Gross National Product for all the people of the country.

Centralized systems amount to a suspension of civil liberties: a limited suspension often unavoidable; one mandated by the law and subject to correction by the voter at election times; but a loss of liberties nonetheless.

Very blunt instruments, in a democratic society centralized systems should be used very sparingly, only when an overwhelming consensus approves, and only where other options are not feasible. When used crudely, centralized systems elicit a rude response.

# 2

## *How Volatile, Energy*

IN THE SUMMER of 1979 in preparation for a long-demanded national inquiry into nuclear power, Canada's federal government sent Harry Swain, a top troubleshooter in the Energy Department, to Toronto. His assignment: to meet with about two dozen of the 300-odd member groups of the anti-nuclear lobby and find out what the fuss was all about. "Why are all you groups getting involved?" he seemed to be asking, bewildered. "Nuclear power has nothing to do with church issues, or any of your social causes. Nuclear power is a technical issue; it needs a technical solution."

To the surprise of the anti-nuclear groups, the government's intention was to inquire only into nuclear reactors. It was now a few months after the reactor accident at Three Mile Island had heightened public concern and shaken the government out of its complacency. By proving nuclear reactors safe, the government reasoned, the issue would be over and the protesters could go home.

Rather than the plaudits he expected, Swain found himself besieged by an opposition alarmed that its concerns were being ignored. Reactor safety is a primary concern only to the towns and cities in the vicinity of nuclear power plants. To the residents of the huge Northern Ontario region, meltdowns and leaks from reactors are theoretical—their concern is the planned disposal of radioactive wastes in their own backyards. The residents of the uranium mining regions of British Columbia, Saskatchewan, and Ontario direct their anger toward the radioactive contamination of water supplies; to the labor unions the concern is over the health and safety of workers in the mines. Farmers like neither the expropriation of their agricultural lands nor the huge transmission towers intruding into their workplace. People in small towns slated for uranium refineries don't like the high incidence of cancers associated with that industry. People living along all the routes to and from nuclear facilities are opposed to the transportation of highly radioactive materials through their communities.

The consumer groups and ratepayers oppose nuclear power because—despite industry claims to the contrary—nuclear electricity has been responsible for dramatic increases in electricity bills. Church groups oppose the sale of nuclear reactors abroad because such sales tend to discriminate against the poor. This is especially true in Third World countries, which lack the extensive electrical grids needed to benefit the non-urban majority. War protestors don't like either the links between the civilian and military nuclear power programs or the spread of nuclear technology to unstable countries like India, Pakistan, and Argentina under the guise that the nuclear fuel will be used only for peaceful purposes. Native groups are furious that their traditional lands are being violated. Doctors are protesting over the health hazards associated with all aspects of nuclear power. Nature conservancy groups see whole river systems (like the Serpent River system in Northern Ontario) being killed by the radioactive wastes dumped in them, while the tourist industry feels the loss of once-lucrative resort areas. Environmental groups, who began early in the 1970s by saying "there must be a better way," oppose nuclear power because they feel they've found the alternative in conservation and renewable forms of energy. Anti-poverty organizations object to nuclear power because there are more jobs in conservation and solar energy. Civil rights groups object because—in all these instances—human rights are often disregarded as the nuclear industry steamrolls over local opposition.

Had the federal troubleshooter recognized that the hostility toward nuclear power drew its intensity not out of a vague fear that a nuclear reactor could explode, like a bomb, but from countless aggravations, dispersed geographically, and affecting members of virtually all constituencies, had he perceived that this hostility was a result less of an abstract fear than the unpleasant sensation of continually having one's toes stepped on, he and his government might then have begun to understand that the protests are not really over nuclear power at all, that nuclear power is only the most visible and most common of the energy causes being protested across the continent.

Hydro-Quebec's James Bay project, the largest hydro-electric project in the world, was bitterly opposed for years. Similarly other hydro-electric sites at Garrison, at the Peace River Valley, at Columbia and Alaska and the Yukon have been and are being fought. Those who think anti-nuclear protesters favor clean, renewable hydro-electricity are often dumbfounded to discover the same people protesting hydro projects.

Arctic pipelines—whether gas or oil—are also fought with a fury hard to understand by those who wonder why one more pipeline is going to make any difference. Why should the caribou herds that might be threatened come before our comforts, they ask? Why should the fear of disturbing something as obscure as the Arctic ecosystem be allowed to hurt the continental economy? The confusion increases when protests accompany the building of oil refineries in New England, tar sands plants in Alberta, oil shale facilities in the U.S. Midwest, oil tanker routes off the West Coast, liquefied-natural-gas ports in New Brunswick and Nova Scotia, coal plants and coal mines everywhere.

When solar space satellites are proposed they will also be objected to, as will the development of fusion energy. These are not signs of inconsistencies among the protesters, not due to troublemaking tendencies, not indicative of an anti-technology bias.

The objections all relate to size. Nuclear power plants not only need to be huge to keep costs down but need a huge support structure extending hundreds or even thousands of miles. Small research reactors of the kind used at universities do not draw hundreds of thousands of protesters. The James Bay project swallowed up an area of land twice the size of Delaware. The damming of small rivers for hydro-electric generation rarely draws more discussion than the rezoning of a piece of land.

At any given time now, there are about one hundred major energy conflicts taking place in North America: farmers toppling transmission towers in Lowry, Minnesota, townspeople up in arms over uranium exploration in Genelle, British Columbia, civil disobedience at Rocky Mountain Flats in Colorado and Darlington in Ontario, government inquiries in California and Saskatchewan. Each conflict is a response to a massive energy project that is in the works, and, as energy shortages become more severe and governments turn to increasingly large, increasingly desperate measures, the conflicts can only get worse.

Except for war measures, nothing approaches energy projects in size and impact. Synthetic fuel plants now start at $6 billion. Arctic pipelines are being quoted at $26 billion. Energy investments have become the single biggest sector of the business economy, so big they are approaching the level of investments in all other business sectors combined. In Canada alone bankers are preparing for energy investments totalling $300 billion over the next decade—sums entirely outside the country's experience. In the U.S., investments into the trillions of dollars are being drawn up. Every dollar spent represents a degree of social disruption.

Huge projects, whatever their character, can't help but step on many toes. In the case of energy much more than size is involved. Nuclear power plants and oil pipelines mean centralized systems—large systems that are controlled at a very few locations by a very few people, making the rest of the population dependent on them. For a country like the USSR—the greatest proponent of nuclear power in the world—the attraction of a centralized electricity grid has always been clear to a government in Moscow exercising control over dozens of different nationalities. Electrification was promoted by Lenin as an absolute necessity from the first days of the Bolshevik Revolution, and every Soviet leader since Lenin—from Stalin down to Brezhnev—has seen the logic of literally being able to pull the switch on any potential uprising.

In North America there is no need for this kind of political control over people. But, as everyone who gets involved in the energy protest movement comes to realize, democracy is what is ultimately at stake. It comes down to a question of power and responsibility being concentrated in few hands.

The farmers from Minnesota who were eventually driven to toppling transmission towers started with some simple concerns. It was rumored that power lines would soon be strung over their properties by their local rural co-operatives, in partnership with large corporate interests. The farmers' concerns began to grow when co-op members who went to some board meetings to find out what was happening were barred from these meetings and were unable to obtain minutes of the proceedings. When they attempted to obtain information through normal channels the farmers were told the information was unavailable. When they tried to exercise their rights at annual meetings of the co-ops they found that the election process had fallen into the control of management. When members asked questions, they were ruled out of order.

Adding to the farmers' frustration was evidence that the closer someone lives to a power line the greater the danger of induced shocks, ion ingestion, non-ionizing radiation, and contamination from ozone and nitric oxides.

After four years of being stymied in corporate board rooms, in the courts, and in the state legislature, of being kept under surveillance, of being the target of propaganda campaigns and dirty tricks, the farmers began taking matters into their own hands. They dismantled towers, shot out thousands of insulators, and partially severed power lines to prevent them from carrying their high-voltage load.

"At all times, we have urged non-violence against people; but not against property," said George Crocker, a Minnesota farmer and Vietnam war veteran, explaining the tactics used since 1977, when, after exhausting legal channels, the civil disobedience began. "When you are in a fight for survival, people have great imaginations. A survey crew arrived on one farmer's property, so he jumped on his tractor which was attached to a manure spreader and let that crew have the load. They left. When company equipment was left parked in a field, five or six farmers got their tractors and front end loaders full of huge rocks and dumped them around the equipment. It didn't move for days. When the company drills holes in the ground for the 190-foot-high towers, they come back next day and find the holes filled in again."

Over one hundred protesters have been arrested in the ongoing attempt to stop the line, which was planned to stretch some 460 miles from a power plant in North Dakota through Minnesota and beyond. The issue long ago stopped being a narrow dispute over transmission towers. The protest has raised fundamental questions that must be asked in a democracy: who should be making decisions, and through what process?

In Canada, secrecy by Crown Corporations like Atomic Energy of Canada Limited (the company that markets nuclear reactors), the Atomic Energy Control Board (the organization that regulates the nuclear industry), and Ontario Hydro (the utility that operates most of Canada's nuclear reactors) extends to whatever these government agencies wish to keep confidential, even the safety studies of their reactors.

Information about the CANDU design was often easier to obtain across the border (because of the U.S. Freedom of Information Act) than from Canadian agencies. Anti-nuclear groups were even driven to ask communist Romania—a prospective purchaser of a CANDU system—for confidential information to which Canadian citizens didn't have access. Even a legislative committee with subpoena powers had difficulty obtaining information it required for its own investigation. The information kept secret was not limited to a few sensitive studies as some members of the committee might have supposed—a full four tons of safety studies were finally released.

Secrecy aside, Canada's nuclear authorities have the same credibility that distinguished the American authorities involved in the Three Mile Island accident; both are given to distortions and half-truths, and, where necessary, outright lies. This lack of credibility is complemented by heavy-handed tactics which provoke outrage in virtually every community the nuclear authorities enter. Communities being asked to store radioactive wastes in their backyards are given public meetings complete with color movies and slide shows showing how safe nuclear waste disposal is. But, should the communities ask that critics of the waste disposal scheme sit on the panel beside the pro-nuclear spokesmen, the government balks.

To defuse the nuclear question, the federal government set up a nominally neutral body called "Committee on Nuclear Issues in the Community," made up of respected academics and public figures on both sides of the debate. Several figures soon resigned from the Committee when they realized its primary purpose was to quietly and dispassionately convince protesting communities of the desirability of nuclear power. In the words of the Committee's co-chairman, Pierre Dansereau, the first to resign, he was being asked to participate in a "whitewash." The Committee phased itself out of existence after it failed bleakly in its task: at its first stormy hearings in Thunder Bay in May of 1978, a well-organized community voted vociferously and almost unanimously against a waste disposal facility. Other efforts by the nuclear industry to find a location to store radioactive wastes have met with similar hostility and failure. The community at large was never consulted by a bureaucracy whose approach was always the same: private negotiations with town councillors followed by a high-pressure public relations campaign discounting any possibility of legitimate concerns. The committee's $200,000 budget was employed not for the communities' edification but for their conversion.

Faced with the likelihood of being the eventual recipients of Canada's first permanent radioactive waste depository, the citizens of Northern Ontario early in 1980 took the initiative: they organized a public three-day conference on waste disposal to fill the void created when the government's committee disbanded in 1979. It was a major undertaking, organized for residents of a geographic area spanning 600 miles east to west, and 600 miles north to south. On February 1, experts in the field of nuclear power, invited from all across North America, converged on the town of Kirkland Lake to freely discuss a matter of the utmost urgency.

At the last minute officials from Ontario Hydro and Atomic Energy of Canada Limited refused to participate, complaining that they had been slighted when the local organizers of the conference had earlier asked them for $2,500 to help with its funding. Other complaints from these crown corporations—that they were expected to participate in a debate and that the conference was "unbalanced" because it presented primarily the anti-nuclear perspective—had been dropped when the organizers agreed to all the government demands for "equity". This still failed to induce the government to forgive the organizers.

After the government's own information program had collapsed, the government expressed great frustration at its inability to resolve the conflicts over nuclear power. Presented with a second chance when the public took up the responsibility, the government seemed less than eager to provide the information it felt the public lacked.

Despite last-minute pleas, the government decided to boycott the conference. In dealing with the public the government adopted not the passive stance of a facilitator but the active one of an opponent. The servant of the people had become their superior, deciding from far-away head offices not to deign to answer their questions, not to dignify their public meetings with the presence of government experts.

People who had come from hundreds of miles away for answers found only frustration. The conference had been transformed from an information forum into a rallying point for northern grievances, radicalizing this region as no amount of anti-nuclear propaganda could have. Local outrage reached such heights that, at a regional government meeting held a few weeks after the conference, delegates representing twenty-two Northern Ontario municipalities voted 43-1 against allowing waste disposal facilities on their territory.

In all such conflicts over energy issues, some huge and distant bureaucracy has tried to impose its will on some smaller entity. Whether it is a creature of big government or of big business does not matter—bureaucracies seem to act the same way. Individual liberties are cast aside for a greater good that was once easy to trust but is now hard to find. Because North Americans waste twice as much energy as Europeans, and because all the Niagara Fallses and other cheap sources of centralized power were long ago discovered, there is not one, large, centralized energy project left on the continent that can't be replaced by small, decentralized energy projects that will provide more energy at much lower cost, far greater speed, and with incomparably less disruption to society. But in the process of replacing large nuclear power plants with small, locally run generating stations, in the process of eliminating Arctic pipelines by better insulating homes, power and control would be necessarily transferred as well: from senior levels of governments to local levels, from large corporations to small businesses and individuals.

The first priority of bureaucracy is self-perpetuation. It is not in the interests of a big bureaucracy to try to achieve its ends through any but centralized means. These centralized means—no matter how well intended—can't help but be dictatorial in direct proportion to their size. For huge utility organizations like Ontario Hydro (whose employees number more than the population of the Yukon) it becomes far easier to force everyone to adapt to Hydro's needs than for it to try to become all things to all people.

Ontario Hydro finds it easier to come up with one solution—nuclear power—and try to apply it to everyone's needs than to come up with thousands of small solutions meeting thousands of separate electricity needs. Nuclear power, it turns out, has needs of its own. Unlike other more flexible ways of generating electricity, nuclear power plants must run flat out all the time in order to run economically. Humans must run flat out too to fit the economies of nuclear power.

But not enough people are working at 3 A.M., and too many people are cooking around 6 P.M. to suit the needs of nuclear power. Ontario Hydro's solution is to try to change people to meet the needs of its machines. The utility intends to penalize those who eat dinner at dinner hour with higher energy costs; it intends to give a bonus to companies that change nine to five production shifts to, say, midnight to 8 A.M. shifts. It intends to do all this to encourage people to change their personal lifestyles and eat, work, and sleep at different times—not out of a nefarious desire to disrupt people's way of life but simply to meet the needs of its nuclear reactors.

Ontario Hydro is a public monopoly. Under a free market system, this kind of interference could never happen—not because smaller enterprises wouldn't try to minimize the costs of their systems but because countless other options—systems cheaper than nuclear power and not favoring one lifestyle over another—would be available for the public to choose between.

Technology is a powerful force. When it escapes democratic control and becomes freed of social restraint, it has only its own needs to fuel. As surprising as this may seem to us, Aldous Huxley understood this long before any nuclear power plants were built. In the foreword to his *Brave New World,* written in 1946, he predicted that

> nuclear energy will be harnessed to industrial uses. The result, pretty obviously, will be a series of economic and social changes unprecedented in rapidity and completeness. All the existing patterns of human life will be disrupted and new patterns will have to be improvised to conform with the nonhuman fact of atomic power. Procrustes in modern dress, the nuclear scientist will prepare the bed on which mankind must lie; and if mankind doesn't fit—well, that will be just too bad for mankind. There will have to be some stretchings and a bit of amputation—the same sort of stretchings and amputations as have been going on ever since applied science really got into its stride, only this time they will be a good deal more drastic than in the past. These far from painless operations will be directed by highly centralized totalitarian governments. Inevitably so; for the immediate future is likely to resemble the immediate past, and in the immediate past rapid technological changes, taking place in a mass-producing economy and among a population predominantly propertyless, have always tended to produce economic and social confusion. To deal with confusion, power has been centralized and

> government control increased. It is probable that all the world's governments will be more or less completely totalitarian even before the harnessing of atomic energy; that they will be totalitarian during and after the harnessing seems almost certain. Only a large scale popular movement toward decentralization and self-help can arrest the present tendency toward statism. At present there is no sign that such a movement will take place.

The "large scale popular movement toward decentralization and self-help" is now well underway. It is taking on public and private monopolies, fighting statism of the left and the right.

At this, Huxley would not have been too surprised. The same battle has been fought before. Neither would he have been surprised at how similarly the battles have been fought, at how little the tactics and strategies have differed over time, at how many of the drawn-out and anguished energy decisions being made today are no more than the predictable echoes of past pronouncements, at how much blind faith continues to accompany technology and at how successfully the questioning of technology has been made an unspoken taboo.

# 3

# *Energy Ideologies*

MOST OF THE fundamental decisions affecting how we live today were arbitrarily made for us fifty and a hundred years ago. The gasoline car won out over the electric car, the diesel bus replaced the electric streetcar, the Greyhound bus beat out the electric passenger train, and the electric locomotive lost out to diesel locomotives and trucks. As a result a revolution took place in the way we live, where we locate our homes, even the air we breathe, shocking our sensibilities yet scarcely penetrating our consciousness.

These were not separate, unrelated contests.

In each case, electrically fucled vehicles were replaced with petroleum-propelled machines. The result is a transportation system that has traded in exhaust-free vehicles for polluting ones, that has rebelled against dependence on overhead wires for the freedom of the congested road. The seemingly unlimited mobility of the car saw the rise of the modern suburb and the decline of agricultural lands, the depopulation and often the destruction of cities. The mass migrations of people out of cities—millions of discrete personal dramas unwittingly directed by an unknowing hand—are being reversed as people spontaneously wake from their suburban dream to opt for urban reality.

Because George Westinghouse won out over Thomas Edison in 1893 for the contract to bring power from Niagara Falls to Buffalo, Westinghouse's alternating current (A.C.) became the dominant form of transmitting electricity. Edison lost out on the direct current (D.C.) system he was promoting and we lost out on a far more efficient technology that, by itself, could have so reduced our dependence on electricity that nuclear power would not yet be needed.

Only recently have we begun to recognize the magnitude of our error. To compensate for the inherent inefficiency of alternating current NASA, followed by American and Japanese firms, have developed a variety of add-on gadgets designed to reclaim A.C.'s loss, and a gradual phasing in of D.C. systems of the kind utilized a century ago may yet come about.

Because Edison decided that large central electricity generating stations—supplying whole cities and towns—were preferable to the home-sized generating systems then in use, he reversed the trend he'd started in 1882 with the installation of a generator in J. P. Morgan's cellar, making Morgan's the first residence to use electric lighting throughout.

The Edison General Electric Company (later known simply as the General Electric Company) shifted the technology, blitzing the continent with city-sized stations and setting a trend that would fix our way of life for a century or more. We have learned the meaning of dependence on distant power sources and long transmission lines, of new words like *brownout* and *blackout.* Our modern style of life is unlike any known to us before. It is dictated by an energy ideology so compelling that it cannot be fully apprehended without understanding its origins. It is an ideology that was and still is proselytized with fervor by secular missionaries disguised in a cloak of science.

"Electricity is an unseen agent at work in our midst," pronounced Canada's federal government in 1976, conjuring up visions of angels among the electrons. "Nobody knows for certain what it really is. It arrives only when summoned and at a moment's notice, since it travels at the speed of light. It is not visible to the naked eye or microscope, yet it has become one of our greatest servants."

This sermon from the state continues with claims that no other forms of energy shall be put before it, as only electricity can fill the every need of all, neither is any need too large or too small for its attention, be it thousands of horsepower, or a thousandth of a single watt. Confidently, we are told electricity is making new converts every day, and present users are deepening their commitment. By the year 2000 we'll be consuming more than five times as much electricity, and by the year 2050, when almost all our energy comes from electricity, we'll be ready to set forth into a heaven on earth.

We have heard that same theme before, more awesomely expressed. "From now on we can travel at the speed of enlightenment or we can dawdle and fool and politick around, and postpone or lose the greatest opportunities man has known," said a former chairman of the American Atomic Energy Commission in 1949. "It is a great privilege to be alive at a time in the world's history when a discovery akin to finding fire or electricity comes along. The sooner men accept that fact and are stimulated by it, the sooner in my opinion will we enjoy the now incredible possibilities of atomic science."

But it took until 1955 for that same organization to give credit where credit's due: "We are living in an era that seems designed to test the courage and faith of free men," a later AEC chairman reverently said. "Yet I do not believe that any great discovery of the atom's magnitude came from man's intelligence alone. A higher intelligence decided that man was ready to receive it."

Government's role as a promoter of electricity has a long history, going back more than fifty years when prominent politicians such as Gifford Pinchot, the governor of Pennsylvania, exulted over the superiority of this form of energy.

Steampower had changed the face of the earth, Pinchot said, and Americans had especially benefited, with the highest use of mechanical power, per person, in the world. With electricity, unimaginable benefits were to be laid before them. Not only would every home have electric light, heat and labor-saving devices and every farm have electricity for milking, feed-cutting, woodsawing and a multitude of other onerous tasks, but electricity would clean cities of smoke and ash, create garden cities out of slums and forge model cottage industries in the countryside. Man would be free to forsake grotesque, gargantuan cities and return back to the land, to his natural home.

Even before Pinchot made his imperfect prophesy, North American man had begun to see the light. Thanks to the organizational genius of the electrical industry, information committees and bureaus were set up after World War I to distribute millions of pieces of pamphlet propaganda to all sorts of influential people. Speakers were promptly provided to chambers of commerce, clubs, and churches. High school budgets got a bonus as free educational literature was made available on all things electrical.

Not until the early 1930s, though, would the full glow of this public-spiritedness shine through. Taking on the mission of "lighting education for the nation" through its Science of Seeing program, G.E. set out to right an injustice that had plagued man since the beginning of time: poor eyesight in infants.

Claiming that poor vision in two million children (out of a total of 10 million) was caused, at least in part, by insufficient lighting, G.E. ran ads titled "One of 10 million children who need LIGHT CONDITIONING." Under a picture of a normal-looking, happy little girl, unwittingly handicapped through the ignorance of her parents, the ad explained that "Light Conditioning is simply providing the right amount of light and the right kind of lighting for eyes at work or play." After a reminder that Light Conditioning is "the science of Better Light for Better Sight," G.E. appealed to the reader to save himself as well as his child by merely calling on the good graces of General Electric and his own good sense: "You can start to light condition your living room . . . today . . . for as little as fifteen or twenty cents. One new Edison MAZDA lamp of the proper size often makes a surprising difference in the amount of light you get for your seeing task. Your electric lighting company has a free light conditioning service. Just phone and a trained Lighting Advisor will measure your lighting with a Light Meter and show you how to light condition your home."

At the height of the program, in the middle and late 1930s, there were more than 2,500 home lighting advisors. Armed with demonstration kits, lighting advisors made calls on homes throughout their sales territories, pointing out changes that could be made at little or no cost to improve existing lamps and fixtures. After a careful scrutiny of dark nooks and shadowy crannies the advisors sat down and filled out comprehensive recommendations for the home: each room was diagnosed and given a lighting prescription, specifying what lamp bulbs, portable lamps, and fixtures were needed to provide cheerful, comfortable, and healthy lighting. The homeowner, of course, was under absolutely no obligation to buy.

Lighting advisors did not limit their calls to private houses. Women's clubs and PTA's were visited. Department store lamp buyers and electrical retailers and wholesalers were called on and advised to stock the best residential lighting equipment. Architects and builders were consulted to aid in the specification of better wiring and lighting in new or remodeled homes.

None of this would have been rational without the research conducted by the Science of Seeing, which found that visual acuity increased by up to 30 percent when lighting levels were increased by a factor of ten, by up to 70 percent when lighting levels were one hundred times as great. At one hundred times greater brightness, eye-muscle fatigue declined, as did nervous muscular tension affecting the entire body. Since human beings are "human seeing machines" that have not only eyes but brains and optic nerves involved in seeing, the entire nervous system and muscular system also benefited from greater lighting levels.

"Seeing work" was much more difficult indoors than under the sun, where lighting levels were hundreds, even thousands of times greater. The result of so much "seeing work" was common eyesight defects (although Science of Seeing scrupulously pointed out that other factors besides lighting could be involved). In addition, glare was found to be an enemy of the eye, along with heavy shadows and harsh contrasts.

To correct these serious deficiencies, the Science of Seeing concluded, we had two options. First, we could change the thousands of systems that were the source of the problem, for example, by printing newspapers on high-quality glossy paper instead of newsprint. Or, we could simply correct the problem with more brightness. Since the lighting industry was well equipped to provide good illumination, Science of Seeing recommended the second option. There was no longer any reason to be handicapped by "barely seeing" when seeing could be made so much easier.

The Illuminating Engineering Society, the standard-setters for the industry, followed G.E.'s leadership and progressively raised minimum lighting requirements. For the benefit of consumers they put tags on lamps manufactured to specifications (three light bulbs per lamp became common). According to G.E., incalculable benefits resulted. Millions of IES lamps were sold. More millions were modernized and fitted with the right size of bulb. New fixtures blossomed in place of old inadequate ones and people learned to appreciate the decorative beauty that light could add to their lives.

The benefits G.E. was bringing to the public were threatened, however, when a Japanese firm introduced a cheaper bulb. To save the public from itself, G.E. introduced the ten-cent bulb, drove the Japanese firm out of business, and then discontinued its own cheap line in the interest of promoting higher standards.

Desirable programs could now be implemented without annoying diversions. The IES brought out detailed standards for school, office, and industrial lighting. The standards, consistently revised upwards and based on the Science of Seeing principles, are taught to this day in North American universities, and are still honored by architects and developers. On the grounds that heavy shadows and harsh contrasts are harmful, all indoor areas have been flooded with light, regardless of the use to which the area is put. Only recently have professionals begun to question these standards. Shadows deemed dangerous and to be avoided at all costs are comparable to those encountered walking under the shade of a tree on a sunny day.

After a forced pause during World War II, the electricity industry decided to bring a little comfort to the Cold War of the 1950s with its Live Better Electrically program—a decade-long barrage of advertising blitzes designed to get us to change our backward lifestyles for modern ones. First there were $100,000 "light for living" Medallion Home contests, $1 million ads in *Life* magazine, "housepower" contests attracting hundreds of thousands of contestants, and well publicized, well glorified annual "Live Better Electrically Women's Conferences" for progressive housewives wanting to better serve their husbands. Then came custom-designed Gold Medallion Homes and twelve-page "Live Better Electrically" ads in *Life.* Finally came the ultimate . . . "Joy of Total Electric Living" campaign. The industry could now point to significant accomplishments: an Electric City of 6,000 Gold Medallion Homes in New Jersey; announcements that even the Seminoles were living better electrically; glowing endorsements from rural electric co-ops which had decided to promote electric living; and the lighting of the Alamo.

The actual improvements in lifestyle were almost too obvious to mention, as the smiles on the Bennetto family faces attested: "Once the Bennettos had a dream . . ." Ontario Hydro's ad informed. "Now electric heating makes the dream come true." In smaller type, under pictures of their dream home, the declared benefits of electric living were: walls "free from film," fewer colds (because "draughts and chilly spots are non-existent"), and enough space saved in the furnace room to provide Mr. Bennetto with a "downstairs office."

Progress proceeded quite satisfactorily. Between 1960 and 1965, the number of electric homes on the continent tripled, passing the two million mark.

Now that the idea of electricity had been well sold, the industry decided we were ready to learn that some sources of electricity were better than others. One form, in particular, was by far the best of all—cheap, clean, and plentiful. "Electricity too cheap to meter" would become an industry byword for the product of nuclear generating plants. Of this there was no dispute. Nuclear power was inevitable, only a matter of time. Any later questioning of nuclear power would be dismissed as heresy.

Hundreds of films praising fission were produced; nuclear classics like *Atom and Eve, Go Fission, The Alchemists' Dream,* and *The Day Tomorrow Began.* Exhibits toured shopping centers and local fairs with elaborate displays showing what life was like before the atom. Walk-in exhibits allowed visitors to see "the atom at work," or to stroll into the core of a simulated breeder reactor to hear recorded pronouncements about its coming and the promise of what lay beyond—the fusion reactor.

To prevent waves of nostalgia over Hiroshima and Nagasaki, G.E.'s advertising had taken pains to stress the acceptability of nuclear power. "Nuclear power makes a good neighbor" ads appeared everywhere, invariably illustrated with nuclear domes in an idyllic setting. Showing a child fishing in the foreground, one ad explained "electric utilities find that clean nuclear power plants play an increasingly important role in their efforts to supply reliable, low cost electricity and that people everywhere welcome them as truly 'good neighbors' in their communities." Then came the nuclear testimonials: "Michigan welcomes nuclear" . . . "Neighbors like nuclear plant in Illinois" . . . "Citizens praise N.J. nuclear plant" . . . "Californians call nuclear power clean, safe" . . . "N.Y. plant helps make progressive city." The texts of the ad backed up the headlines. The mayor of Oswego, N.Y., boasted, "It will enhance our reputation as a progressive city and a clean, modern, good place to live." The residents of Eureka, California, another ad claimed, "are urging us to build another nuclear power plant in their community."

And in G.E.'s ad for its Big Rock Point station, they proudly passed on the feelings of the local utility company: "We are operating a plant that is safe for anybody—next to the plant or ten miles away. The utility is convinced that nuclear power plants are a welcome addition in any city—even downtown."

Toronto may become the first city to make G.E.'s prophesy come true. The just-completed St. Lawrence housing complex in the city's downtown area is being eyed by Canadian nuclear researchers as a good future prospect for installing a small-scale nuclear district heating plant.

By the 1970s, the nuclear dream had been sold, and advertising shifted to the operating performance record of nuclear power plants. Ontario Hydro's Pickering plant (which had been proudly introduced to us as "a bouncing baby") was now vigorously promoted as having the best performance of any nuclear plant in the world . . . but it had much competition among other utilities which managed—despite difficulties—to keep their plants operating most of the time too. One utility, particularly proud of its economic accomplishments, advertised "an enviable record for any power plant" while going out of its way to scoff at nuclear critics: "You've heard about the real and imagined ills of nuclear power plants. Some of it's true; a lot isn't. We'd like you to hear about one of those nuclear plants that doesn't make the news . . . a smooth running station."

This station would later receive hundreds of millions of dollars in unpaid advertising. It is the one at Three Mile Island, near Harrisburg, Pennsylvania, and it set nuclear advertising back fifteen years. Once again, reassuring ads would have to be drawn up. A full-page effort in the *New York Times* had Edward Teller, the father of the American hydrogen bomb, declaring "I was the only casualty of Three Mile Island." The hysteria of those who thought the Harrisburg accident cause for concern apparently so exasperated Teller he needed to be hospitalized. And Ontario Hydro, in an extensive newspaper, television, and radio campaign, invited one and all to pack a lunch and bring their families for "a nuclear picnic" at Pickering.

Three Mile Island also brought to a height the grammatical inventiveness of the nuclear power industry, later recognized by the National Council of Teachers of English. In awarding their annual Doublespeak award of 1979, William Lutz, an English professor at

Rutgers University and chairman of the Doublespeak committee, acknowledged that nuclear spokesmen outdistanced all contenders in a tough year by inventing "a whole lexicon of jargon and euphemisms." The nuclear accident itself was, in the words of industry spokesmen, referred to as either an "abnormal evolution" or a "normal aberration." Rather than talk of "fires," the term "rapid oxidation" was used. "Explosion" was described as "energetic disassembly." And instead of stating that plutonium had contaminated the reactor vessel we were informed that the radioactive element had "taken up residence" there.

Nuclear ad campaigns in the U.S. had slackened by the late 1970s as private utilities began finding that nuclear—despite its magical appeal—was becoming too expensive to meter. But in Canada, where government subsidies to the nuclear industry cushioned the need to be conscious of making sound economic decisions, the nuclear dream could go on undisturbed. Ontario Hydro spent $700,000 furbishing its nuclear communications centre at Pickering—a mini Disneyland of nuclear delights—and about $500,000 per year running it. The utility's public relations department—among the largest on the continent with a staff over one hundred—continued to crank out free literature proclaiming its immunity from nuclear accidents into the 1980s. And the growing proportion of nuclear electricity in Ontario (where hydro-electric and coal plants are shut down or mothballed to make way for them) is proudly pegged at 33 percent, rising to 50 percent.

Over the last half century, industry has put on its best face for the public. Its information, for the most part, has been accepted as factual. But the industry has a second face as well, which decides what information the public gets to see.

# 4

## *Power Brokers*

THE FIRST LONG-DISTANCE transmission of electric power, from Niagara Falls to Buffalo, was financed by J. P. Morgan. A decade earlier, after Edison succeeded in lighting up New York (the first city to be electrified), it was to Morgan's office that Edison went as, according to a newspaper report, "throughout a third of the downtown district little lamps began to glow."

Morgan's office, like his home, had already installed electric lighting. As an early backer of Edison, putting up much of the $500,000 Edison used to develop his immensely successful light and power system, the visit was more than symbolic. Electricity had become big business. The Edison Electric Illuminating Company of New York, which in 1882 had just installed the first central generating station linking fifty-nine customers with fourteen miles of underground cables, was accompanied that same year by the creation of the Canadian Edison Manufacturing Company to sell central generating stations in Canada. By 1887, sixty Edison central station companies had been established. In 1889 the Edison General Electric Company was formed. When it merged with another giant in 1892 to become the General Electric Company, the new enterprise became the undisputed industry leader, with 1,277 central stations supplying electricity to two-and-a-half million incandescent lights and 110,000 arc lamps. Royalties alone from the central station companies amounted to nearly $2 million per year, and G.E. was also serving 435 electric railway companies with 8,836 trolley cars rolling over 4,927 miles of track. The company was building

1,600-horsepower electric locomotives and manufacturing equipment for an elevated electric railroad six miles long for the Chicago World's Fair of 1893.

Morgan, the man who controlled this corporate empire, had come by his interest in electricity honestly. As a consequence of controlling the New England railroads, Morgan would gain mastery over the mining enterprises so dependent on rail transportation. INCO and U.S. Steel were (in the language of the day) "Morganized." Complementing them was the colossal coal industry that would fire his electric generating system. Morgan's Reading Railroad alone owned at least half the hard coal in Pennsylvania. When "white coal" (waterpower) became popular in generating electricity, Morgan emerged with control over nearly 50 percent of all the commercial waterpower developed within the U.S.

To find markets for his electricity Morgan acquired Western Union (America's first great industrial monopoly), AT&T (incorporated in 1885), electric railroads like the New York, New Haven and Hartford, and electricity-dependent entertainment like the motion picture and radio industries. Through G.E., Morgan controlled RCA, which in turn controlled RKO and NBC. AT&T owned Bell Laboratories and Western Electric. Keith-Albee-Orpheum Corporation, the motion picture distributing giant, was founded by RCA.

To Morgan, these were not unrelated enterprises but all part of the same industry, a "community of interests" in the electricity business which grew in a logical fashion, always toward the legitimate goal of maximizing profits. Initially, central generating stations started spinning at sundown when lighting was required, eased up after midnight and closed when the sun rose. Street railroads were required not only for the profits to be made from them directly but also to increase the efficiency of the central generating stations by keeping them busy in the daytime.

For marketing reasons, central generating stations, rather than individual units, became the norm. Morgan realized it would be far easier to convince prominent businessmen to set up a local light and electric company to service thousands of customers than to wait until everyone in a community saw the benefits of electricity.

Municipalities would be only too eager to grant local monopolies to electricity producers, and G.E. would see to the building of the plants and the collection of royalties for their use. Through licensing arrangements, General Electric would be used for replacement parts and expansion of facilities. Unlike numerous small customers, the local utilities would not need to be resold on their products when old generators outlived their useful lives and needed to be replaced.

The logic of Westinghouse's A.C. led to Edison's exit from the power business. Technological merits aside, Edison's preoccupation with D.C. had become a losing proposition since a French syndicate had succeeded in cornering the world's copper market. Copper prices doubled. The D.C. system of transmitting electricity, which used a great deal of copper, was being priced out of the market. Rather than fight the French monopoly, Morgan decided to fight Westinghouse instead by dumping Edison and switching to A.C. The attack was repelled by Westinghouse, but the message was not lost. In 1896 Westinghouse entered into a patent-sharing agreement with G.E., agreeing to 37½ percent of their patent profits to G.E.'s 62½ percent. That same year G.E. and others organized the Incandescent Lamp Manufacturing Association and between 1901 and 1906 G.E. set up the National Electric Lamp Company to secretly buy out other companies with National's own stock. Over twenty companies were absorbed into the federation, which changed its name in 1906 to NELA—The National Electric Lamp Association.

In an era of great trusts and combines, the electrical industry was one of the fastest to become monopolized. The patent-sharing agreements effectively prohibited new products from being brought onto the market by controlling access to all the patents that might be needed in the formation of new products. Though a Federal Court decision in 1911 held that G.E. and others violated the Sherman Anti-Trust Act, the electric monopoly would survive. Innovations from the industry would end despite G.E.'s motto that "progress is our most important product." Electric toasters, electric ranges, electric refrigerators, electric dishwashers, vacuum cleaners, washing machines, steam irons, deep freezers—all were developed by entrepreneurs. According to T. K. Quinn, a former vice-president of G.E., the inability of large corporations—including his own—to innovate is not the exception but the rule.

"I know of no original product invention, not even electric shavers or hearing aids, made by any of the giant laboratories or corporations, with the possible exception of the household garbage grinder, developed not by the research laboratory but by the engineering department of General Electric. . .the record of the giants is one of moving in, buying out, and absorbing the smaller concerns."

G.E. would proceed to fix prices and divide markets around the world. The production end of the industry had already been barred to newcomers. To lock up the distribution end, G.E. developed the "agency system" (an approach soon followed by Westinghouse) to control retailers who might want to sell competing brands.

Though investigated by the Federal Trade Commission in 1919 and a committee of the New York State legislature in 1922, though taken to court in 1924 and 1925, G.E. successfully frustrated government plans to regulate the electrical industry.

The industry's full political clout would not surface until 1928, when the United States Senate decided to embark on a full investigation of the financial dealings of the utilities. What emerged to oppose the investigation was, in the words of a senator from Wisconsin, "the most powerful lobby in the history of the nation." Senator Walsh of Montana, who headed the bipartisan investigation, concurred, calling it "the most formidable lobby ever brought together, in my time at least, for fifteen years, representing capital to the amount of nearly $100,000,000,000 and representing what? The general public, the consumers of electric energy and the purchasers of securities that are put out by these companies? Not at all; but representing the companies to be investigated."

The legacy of this lobby remains with us today. Its electrical empire continues to dictate our lifestyles; it became so entrenched that alternative means of generating electricity have yet to be established. The supremacy of the present electrical system was maintained, above all else, through the prudent dissemination of information and a rare co-operation among competitors.

G.E. and the electrical lobby effectively controlled all phases of electrical production, from the raw resources needed to generate energy to the finished products that consumed the electricity—a conflict-of-interest situation that needed to be maintained with care.

One agreement between utilities and lightbulb manufacturers that "advances in the lighting art should not be at the expense of wattage," was necessary to protect utilities from energy-saving innovations like the fluorescent bulb. Were they to be marketed commercially before being perfected to consume more electricity, utilities would face reduced sales of electricity. G.E. took pains for itself to ensure new light bulbs did not burn longer: utilities expected G.E. to extend them the same courtesy by ensuring the bulbs also not burn less power.

Carelessly, G.E. gave the public a preview of the fluorescent's efficiency at no less prominent a place than the New York World's Fair. Complaining G.E. was "violat[ing] the spirit of the understanding that our group had in Cleveland" with the fluorescent display, a utility letter expressed pique that "20 watts of fluorescent lighting are compared with 20 watts of incandescent lighting, the sign purporting to read something to the effect 'See the difference between equal wattages of fluorescent and (ordinary) lighting.' Of course, the readings . . . show dramatic differences."

The display was removed. G.E. had been uncharacteristically remiss. Since 1919, when the utilities launched in Illinois the prototype of future public relations operations, the public never had cause to feel neglected. By 1921, five million pieces of literature had been distributed. Explained B. J. Mullaney, the utility public relations director: "Those five million pieces of literature, all helpful to the utility industry, were not merely scattered broadcast, but were definitely placed: With newspaper editors for themselves and their readers; with customers of public utilities; with businessmen, bankers, lawyers, employers (for their employees), teachers, preachers, librarians, students in colleges and high schools, mayors, members of city councils and village boards, public officials of all kinds, and candidates for public office. Members of the legislature, for example, received informative matter on public utility questions not after they were elected, but before they were even nominated."

The public relations effort did not stop there.

"A news service goes regularly to the 900 newspapers in the State, about 150 of them dailies. Speaker's bulletins are issued . . . The bulletins furnish ample material to any intelligent person for sound talks on each subject and they have been widely used. A bureau is operated to find engagements, before clubs, civic associations and so on, for dependable speakers. . . .

"Pertinent addresses and articles by important men, resolutions or other expressions by chambers of commerce and other bodies, exceptional editorials and the like, and special matters for customers, investors and employees have been printed and circulated among special classes by hundreds of thousands. More than 800 Illinois high schools are regularly furnished informative literature for classroom theme work, and debating-society use."

The public relations scheme's success surpassed all expectations. Within one more year, P.R. programs were underway in half the states of the Union, "functioning in such a way that newspapers in six states are using an average of 8,500 column-inches of material every month of which approximately one-fourth or about two-thirds [sic] is editorial comments favorable to public utilities. That means something more than 500 pages of reading matter every month . . . with a total circulation running between two and one-half and three millions of readers . . ."

By the end of 1922, public relations programs had been organized in a majority of the states, the U.S. was divided into twelve zones or geographic divisions of the National Electric Light Association, and in the Association's annual proceedings it was declared that "the one supreme danger that threatens the permanency and credit of our industry is the dissemination of false statements or erroneous information by misinformed or ambitious demagogues. To overcome popular misconceptions, prejudice, and error, we must publish the truth and circulate it widely."

And circulate it they did, leaving no room for editors to misunderstand. The stories were sent already written, laid out, and prepared to be put on the press: "Please . . . deliver the enclosed newspaper story to Mr. Boney with instructions for him to use it as he sees fit. This story is being sent this week to 112 weekly newspapers in Alabama. It is going out in plate form already headed, and to be run as strictly news matter."

Nothing pleased the utilities more than their discovery of the wire services, which were able to take much of the burden off NELA's Public Utility Information Committees: "Frequently the large press associations, through which the newspapers get much of their news, found the material issued by the committee of such importance and timely interest that they themselves sent much of it directly to the newspapers," one committee reported.

The wire services also were dependable. When NELA's director of publicity wanted a highly critical report of public ownership in Ontario to get the widest possible circulation, he "succeeded in getting stories carried by the Associated Press, United Press and Hearst Service. By so doing we believe we will cover the country pretty well so far as the larger newspapers are concerned. Of course the article will be comparatively short, the longest being that carried by the Associated Press, which has agreed to send out 1500 words of an early morning telegraph story. . . ."

Of the three services, Associated Press was the most co-operative. "The Associated Press sends out practically everything we give them," wrote one executive. "Whenever we have had occasion to use the Associated Press our material has gone over with a batting average of 1.000 . . . The Associated Press here has often taken stories from the Utilogram, our news bulletin, and sent them throughout the State without any suggestion from this office," wrote the director of the Kansas Public Service Association. Other directors sent in correspondingly glowing reports. The industry "truthhoods" were being spread far and wide.

By 1924, the importance of proper information had become indisputable: "Fundamentally the prosperity of our business, its growth and success, are built upon a proper state of public mind without which we do not get the money with which to build plants, and we do not get the favorable reaction from the public mind that enables us to sell our product to the best advantage and at fair rates after we have produced it." Also indisputable was the need for more of it: "As the business expands the need for cultivating this public attitude becomes all the more acute. The whole future of the industry rests upon our ability to continue the favorable mental attitude toward the creation of which we have made a good beginning."

Public relations, in fact, had become the primary business. Said Matthew Sloan, President of the Brooklyn Edison Company and that year's Chairman of the Public Relations Section of the National Electric Light Association, "I weigh my words carefully when I say that I believe the work with which this section is charged is the most important in the whole broad scope of activities of the electric utilities." H. T. Sands of the giant Electric Bond and Share Company and the previous year's chairman punctuated his successor's remarks: "The commercial phases of this proposition sink into insignificance when compared to the public relations possibilities." Another utility president declared, "I believe we will all agree that [public relations] yields place to none . . . in the matter of valuable service to us all."

And so the industry decided to get serious about the business of educating the public. It went straight to the colleges and other institutions of learning. Said NELA's Managing Director soon after to a utility convention: "I would advise any manager who lives in a community where there is a college to get the professor of economics, let us say—the engineering professor will be interested anyway—in your problems. Have him lecture on your subject to his classes. Once in a while it will pay you to take such men and give them a retainer of one or two hundred dollars per year for the privilege of letting you study and consult with them. For how in heaven's name can we do anything in the schools of this country with the young people growing up, if we have not first sold the idea of education to the college professor?"

But the "most important work," according to the Committee on Cooperation with Educational Institutions, "will be outside of the engineering schools. It is desired that coming generations of bankers, lawyers, journalists, legislators, public officials, and the plain ordinary 'man in the streets' shall have an intelligent and sympathetic understanding of the peculiar conditions under which utilities operate."

Though this education policy would expose the industry to an occasional charge of propaganda, the real propagandists, as the industry well knew, were the critics of the electrical industry. A utility survey uncovered indisputable proof that even schoolbooks were being used for propaganda purposes.

In Texas: "All the textbooks used in the schools are more or less erroneous on the utilities and generally on the fundamentals of economics," reported the director of the Texas Public Service Information Bureau. A study of Illinois found "most of the textbooks as relate to the public utilities industry . . . no less than poisonous." In Missouri, all school textbooks were determined to be "wholly valueless, and in many instances poisonous. As a matter of fact, 97 percent of the textbooks used in the public schools affecting public utilities are written by socialists and advocates of public ownership."

Drastic action was called for. In its annual convention of 1925, NELA's Public Relations Chairman Sloan, saddened that "it is perhaps impossible to make our public relations work so inclusive that it will stretch from the cradle to the grave," exhorted his colleagues to "at least begin early with it," noting that "there is a particular need for furnishing correct information about our industry in the schools. School books in wide use all over the country have recently been analyzed. Many of them contain startling misstatements about public utilities. The pupil studying such material, hearing it discussed in the classroom, starts life with a warped and biased point of view regarding public utilities, and this point of view formed in the impressionistic years of youth is only too likely to remain unsympathetic and antagonistic through all future years. It is high time to be busy in furnishing correct information to the pupils in our schools wherever and whenever it can properly be done."

Above all others, the weight of this responsibility would fall on John B. Sheridan, Secretary of the Missouri Committee on Public Utility Information. "Give us the child at 7 years old and we care not who educates him thereafter, he will be ours," he declared. And then "Teachers come and go. Textbooks remain. The text taught is more important than the teacher."

A Sub-Committee on Textbooks of the Educational Committee of the National Electric Light Association was formed, with Sheridan as its Chairman. He became the supreme judge over textbooks affecting public utilities, and campaign manager for the "revision of textbooks on civics, economics and civil government."

Waging a successful campaign required moderation and tact. Sheridan wrote to one utility that "great care must be used to avoid going too far since if the public were to get the idea that textbooks were being used as propaganda for public utility companies the reaction would be worse than the original misinformation." Another utility was advised, "the matter is an extraordinarily delicate one and may be handled by tactful personal contact." The following formula was suggested: "My idea was that we should go direct to the textbook writers, tell them that their books were all right for the period for which they were written, but that changes in the past ten years make a new textbook necessary. This will give the boys a chance to write new textbooks and make some more money, and we can show them individually just where-in the old textbooks are obsolete."

The formula produced the proper results, and books were soon produced to utility specifications.

Success was also recorded in placing sound books in libraries. "Where textbooks which were grossly unfair . . . were used in the high schools, we took the matter up personally with school officials, either through local managers or directly," Iowa wrote. "In nearly every instance where such textbooks were used they were removed and placed on the library shelves for use as reference matter only." The Rocky Mountain Committee wrote: "if necessary we will go to considerable expense . . . to carry out the plan now being formulated for placing text and reference books in virtually every university, high school and public library in Colorado, New Mexico and Wyoming."

But the greatest effort went into placing the industry's own publications in the schools "to fix the truth about the utilities in the young person's mind before incorrect notions become fixed there."

"Working the schools," the Texan Committee on Public Utility Information was told, "is something which must be handled very carefully. A misplay in that will do a lot of harm. You cannot afford to let the public think or the politicians to come out and say that the electrical light and power companies, water, gas and telephone companies are trying to circulate propaganda through the public schools. Your approach is through the superintendent of schools in each city . . . in the form of pamphlets which are put in their hands merely for the purpose of giving them proper information."

The pamphlets were popular beyond all expectations, "completely exhausting the supply." In Missouri, 659 (out of a total of 790) high schools used the publications; in Illinois it was 672 schools, in Ohio 553 schools. The route to the minds of youths had been found, and now the danger was in limiting the effort to the schools. The Boy Scouts were first recommended as "the surest, quickest way, and most effective antidote for radicalism in America . . . The general manager should identify himself with the Boy Scout movement . . . and should encourage some of his lieutenants to become scout executives, scout masters, etc. . . . The knowledge of this good work . . . soon finds its way into every home and is deeply appreciated by all boys and their parents, hence the benefit that can accrue to the public utility company through the activity of its officials . . . similar good work can be done through . . . Girl Scouts."

Leaving no stone unturned, the industry then went after the kindergarten set, publishing a "thirty-two page book, printed in color, for children. It is entitled 'The Ohm Queen,' and is intended to tell the story of electrical service in the home, particularly to the young people who are such an important element in our homes, and who will be the customers, the investors, the voters, and the law-makers of the future." The first printing totalled 400,000 copies.

Meanwhile, the industry had come to realize that what was good for the child was good for the mother. "It is about time that we awoke to the fact that through women's clubs and through the cultivation of the women in the women's clubs, we have one of the greatest avenues for dissemination of correct information relative to the public utility and of nullifying incorrect information," one industry leader said. The best means of communication with groups of women was found to be "the holding of company 'at homes' and teas." A single company entertained 10,000 women this way, the function having been "arranged, managed and presided over by members of the women's committee." Well-known women were paid to write articles, in collaboration with utilities, which were then placed in some of the leading magazines in the country.

Through $80,000 secretly given over a period of three years to the General Federation of Women's Clubs, the federated clubs of thirty-two states carried on campaigns for the electrical industry.

In the end the education of the public included everything and occurred everywhere: in the universities and colleges, in high schools and grade schools, men's and women's organizations of all kinds, the public platform and the daily and weekly press, in radio, motion picture, music, and drama. When asked by the Federal Trade Commission whether any form of publicity had been neglected, NELA's Director of Public Information replied, "Only one and that is sky writing."

But though millions upon millions of pieces of information had been distributed, though every means available was being used to promote that special public utility point of view, much more than their narrow self-interest was at stake. As they saw it, they were fighting not just for the shallow right to make great profits but for the basic fundamental right to make even greater profits. The fight, as the industry often needed to remind itself, was really over democracy, the real enemy being the socialists who advanced public ownership of electrical utilities. Corporate warfare between public and private utilities lasted more than a generation. When the smoke finally cleared, the public and private utilities came to a revolutionary but counter-intuitive discovery. They had been fighting the wrong foe all along. Though the managements of public and private corporations report to different kinds of shareholders, in practice the distinction is artificial. The best measure of any management's success is its ability to grow. In this, public and private utilities found no conflict, co-operating among themselves just as the various private corporations had done and continue to do. Public ownership and private ownership eventually assumed the status of meaningless labels. But first there was the tortuous period in which the utilities believed their own propaganda.

# 5

# *The Socialist Menace to the North*

THE TITLES OF the articles in utility journals in the 1920's and 1930's indicated a certain pre-occupation:

*Socialism Means Death to Initiative*
*Government Ownership Advocates Shade from Deepest Red to Mauve*
*Municipal Plants Fail*
*Russia Tried It Too*
*Taxpayers Are Asked to Buy Utilities for Politicians*
*Government Control Fails, Russ Chiefs Admit*
*Europe's Government-Owned Phones Silent at Night*
*Government Ownership Would Increase Taxes by Billion*
*High Cost of Living Noticeable in Cities That Operate Utilities*
*Government Interference in Business a Growing Menace*
*It Was a Failure—Municipal Ownership Plant Given Away*
*John Spargo Was a Socialist*
*Municipal Ownership Is Biggest Drag Civilization Has About Its Neck, Says Mayor*
*Editor Flays Tendency to Socialize Business*
*Public Ownership in 700 American Cities Fails*
*Italy Awaking, Rejects Socialism*
*Would Plunge Utilities Into Politics*
*Municipalities Selling Plants*
*Six Hundred and Sixty City-Owned Plants Abandoned*

For those who failed to grasp the drift of the industry articles, their theme was spelled out by the Public Policy Committee of NELA at its 1926 convention: "At the present time, the major policy of our industry has to do with socialism. Like the single tax, it is an ever present menace to our people. So completely has this policy been discredited that almost never nowadays does the advocate of a socialistic scheme admit the name."

The Public Policy Committee did not explain why the industry's major effort was directed against a social theory so completely discredited, and the industry did not need to—there was no dispute. "An attack upon the principles for which we stand is an attack upon our government itself," another NELA spokesman declared, echoing the patriotic resolve of the previous annual convention, where it was solemnly declared: "There can be no higher service than the preservation of republican institutions . . . It is just as much the business of the electric power and light industry to preserve these republican institutions as it is for it to give its patrons service and to make profits for its shareholders . . . The country cannot exist half socialist and half free . . ."

The next year, 1927, the industry took to the air waves, revealing a socialist conspiracy on the forty-station radio hookup celebrating "Electric Night"—the forty-eighth anniversary of Edison's invention of the incandescent lamp. "When we think of the freedom that is given to every man, woman and child in America to develop the spark of divinity with which he is endowed . . . it is almost unbelievable that there could be found those who would tear down this marvellous system. And yet there are today subversive movements at work in our land, fathered and fostered by those who . . . would foist on America the shockingly, brutally lowering system of certain backward civilizations. Today they are marshalling their forces in an attempt to put the Government into the electric business . . . How long, my friends, will it be, after the electric business is stifled before the same principles will be applied to the lumber business . . . and the coal business and the grocery business and every business, for that matter . . ."

The concerns about the impending communist takeover reached fever pitch as utility executives took to the hustings to sound the alarms: "None but the blind, deaf and dumb are justified in doubting the existence of efforts, insidious and deliberately promoted, to change the intrinsic character of our government, and eventually its form . . . It [the subversionists] is a troublesome, not to say dangerous, minority because of its composition and the insidious way in which it works. Some of the minority are communists of the deepest Russian 'red'; others are a little socialist 'red,' shading into 'parlor pink'; still others come in delicate mauve tints and call themselves 'progressives.' And they all pull together when 'government and business' are at issue . . . Revolution, even to complete destruction of our form of government is their objective."

To counter the omnipresent but imaginary communist foes in elections, the following formula—still in use—for fighting against great odds was given to the corporate candidates for political office: Do not try logic or reason, but try to pin the Bolshevik label on your opponent.

For the most part, the industry's attack on socialism was couched in abstract theoretical language, or took the form of an attack on distant countries like Russia. When threatened directly, the industry's patriotism turned more practical.

Nothing threatened the American electrical utility business more than Ontario Hydro, the socialist menace to the north, the only major publicly owned utility on the continent and the model for many major projects proposed for the United States. It was big, it was visionary, its customers were happy with its service, its electricity rates were low, it was a pioneer in rural electrification, it was beyond the scandal of corporate manipulations. Ontario Hydro worked, and if the major electrical projects being proposed in the U.S. were to stay out of government control, Ontario Hydro had to work. Or Ontario Hydro had to appear to not work.

HERE IS THE TRUTH ABOUT ONTARIO HYDRO-ELECTRIC blared the huge sixty-point type across eight columns of the Boston *Herald*—a full page Sunday feature that would be reprinted in other newspapers and reproduced by the thousands for later mass distribution.

The article was representative of the industry's general approach to presenting information. Written by a NELA employee, who quoted as authorities other NELA-subsidized experts, the article purported all the while to be an even-handed inquiry presenting both sides of the public ownership issue: "Is This Model of Government

Ownership Run as Economically and Efficiently as Private Ownership Could Do It?" the eight-column sub-headline began. "Answer is Provided in Comparison of Conclusions Reached by Two Opposing Experts, with Enlightening Figures."

But the article did describe the leading cast of characters in the raging controversy over Ontario Hydro including: "James Mavor, Emeritus Professor of Political Economy at the University of Toronto. His book is called *Niagara in Politics* and constitutes a rather thrilling story of political manipulation." "Professor E. A. Stewart, head of the Department of agricultural engineering, University of Minnesota, has also made an important contribution to the known facts about Hydro. His personal investigation absolutely disproves the widely circulated statements by Senator George W. Norris and others that the farmers of Ontario are enjoying such a great boon through cheap electric current."

Mavor, whose book was dedicated to a "discussion of the failure of public ownership in Ontario, and the insidious methods of politicians" was on NELA's payroll at the time the book was being written. NELA's dread of state ownership was presented forthrightly as its *raison d'être,* the book declaring that the author "believes that government operation is a dangerous and destructive fallacy and has written this book to prove it in the instance of the Ontario Hydro Electric Commission."

Stewart, who was sponsored by NELA to conduct his impartial, scientific, and scholarly survey, claimed his findings were studied and confirmed by Ontario Hydro itself, as he needed to reconfirm time and time again. "My data were all checked by [Hydro's] engineers before being published," he wrote a disbelieving editor.

Stewart was only half right. His findings were studied by Hydro, but far from confirmed—he had left Toronto before the check could be completed. "Changes and additions to report as requested by Commission's engineers were never made by Stewart," reported Ontario Hydro's chairman. "His original incorrect report was printed and his excuse for doing same was that the corrections did not arrive in time to be included in his publication. Not only are figures published in Stewart's report incorrect in many instances, but statements throughout the report are not in accordance with facts."

Quoted in the Boston *Herald* article was Carl Thompson, hated by the utilities as America's leading advocate of public ownership, the Secretary of the Public Ownership League of America, author of *Public Ownership* and chief publicist of the Ontario Hydro system—rarely failing to note the contrast between Hydro's low rates and those charged by private utilities.

Though NELA's John B. Sheridan went on record saying the best way to treat Thompson "would have been to throw him in a ditch and to hell with him," the industry genteelly confined its activities to harassing Thompson with swarms of private utility hirelings, having his meetings boycotted and his addresses abruptly cancelled.

Often, the only sections of Thompson's speeches objected to by the industry were those dealing with Ontario Hydro rates. "Reading of a verbatim report of the address made by Thompson at Charles City, Iowa, revealed but three short paragraphs that could be called objectionable," said a utility report. "These short paragraphs dealt with comparisons of rates charged by the Ontario Hydro-Electric Commission of Ontario, Canada, and rates charged by privately owned utilities."

But this was enough of a provocation to prompt NELA to orchestrate a public outpouring of protest letters at places Thompson was scheduled to speak.

From P. B. Linville, of the Bank of Edina, Edina: "We have before us the past record of Mr. Carl D. Thompson who will, as billed upon your program, talk to the people of this vicinity . . . Might we offer a protest against this man being heard from your platform in this place? While we appreciate that you at all times endeavor to give a diversified program, we feel that a socialist of this type, undoubtedly of the most dangerous, should not appear and spread his propaganda in our peaceful community . . ."

From F. R. Schofield, Editor of the Edina *Sentinel:* "We are opposed to any man who is a socialist appearing . . . to spread his doctrine; likewise we are opposed to any man advocating Government ownership of anything on our platform . . . This statement is made for present objection to the name of Carl D. Thompson, a socialist of Chicago, but we wish it to be distinctly understood it goes for all others. Positively do we object to any spread of such mental diseases . . ."

These protests and others did the trick, and more suitable arrangements were made as the local manager of the power company at

Edina, Missouri, wrote to the Missouri Committee on Utility Information: "Regarding the Carl D. Thompson matter upon which we sent you some reports and letters recently, I am glad to advise you that we were successful in preventing his speaking in Edina last night as scheduled. He was in town, but did not appear on the program. His time on the program was taken up by the entertainers who were here that afternoon, and it was announced from the platform that his engagement had been canceled by request."

The industry deserved full marks for its thoroughness, resourcefulness, and persistence. Talk of Ontario Hydro and public ownership was hunted down and stamped out. The industry's efforts were of a continental scale, sweeping in both their aims and their accomplishments.

These efforts staved off public utilities, but only until 1933, when over furious protests, Franklin Delano Roosevelt created the Tennessee Valley Authority, modelled largely on Ontario Hydro. As Governor of neighboring New York State, F.D.R. had been a keen witness to the remarkable history of the Canadian utility. He had no sign more promising that the TVA could survive the onslaught of the private electrical interests than the fact Ontario Hydro had succeeded, against formidable odds, and against the very same interests.

Ironically, without J. P. Morgan and everything he represented, Ontario Hydro might never have been formed. It was Morgan who first taught Ontarians the meaning of an energy shortage when, due to labor strikes around the turn of the century, an embargo was placed on the coal that Ontario imported from the States. Southern Ontario's wood had already been depleted and coal, which had cost Toronto $3.50 a ton, now cost two and three times that, with imported coal from Wales costing $10 a ton.

Public sentiment in the U.S. was against the coal barons under Morgan's leadership. Miners had been working under extremely dangerous conditions for extremely poor pay.

The union offered arbitration. The coal barons refused. Pressure on Morgan to intervene produced no results. An organization of businessmen in the coal regions appealed to President Theodore Roosevelt to compel arbitration: "Is J. Pierpont Morgan greater than the people? Is he mightier than the government? . . . Morgan has placed a ban upon us which means universal ruin, destitution, riot and bloodshed . . . We appeal from the king of trusts to the President of the people."

Six months into the strike, the coal shortage was severe, with widespread suffering among the poor in cities. Still the coal barons refused arbitration. Disregarding appeals from the governor and attorney-general of Pennsylvania and the president of the United States, Morgan refused to talk until the miners called off the strike.

Energy-starved Ontario missed none of this. The abuses of the great monopolies and trusts of the era did not need elaboration or translation for the Canadian economy.

Through the American Electric and Illuminating Company Morgan had been selling generators in Canada since 1878 and in 1879 Edison had come to Montreal to produce and install the first Canadian incandescent light in a factory on Craig Street. Electrical developments spread quickly after that—by 1890 there was hardly an Ontario village of over three thousand inhabitants without an electric light station of some kind in operation, and few of the important towns in the provinces were without electric lighting. Canada at that time had 13,530 arc lights and 70,765 incandescent lamps and, almost invariably, Morgan was somehow involved.

The individual members of the leading electrical syndicate in Canada were tightly linked to General Electric (one syndicate member later becoming Canadian General Electric). After J. P. Morgan arranged to tap the American half of Niagara Falls, the syndicate in Toronto became interested in controlling the Canadian half, but like other private interests who had tried before them, they found the venture too speculative to proceed. That wouldn't stop them from trying to stop others.

Out of Ontario's desperation for a secure source of energy, out of fear of having its economy eclipsed by the industrial states to the south, out of loathing for the energy monopolies which were interfering in everyone's lives, the Hydro movement was born. It was a municipal movement with a great deal of grass-roots support headed by the Mayor of London, Ontario, Adam Beck, and backed by other mayors and small business interests. They were up against the provincial government and the big business interests, but by 1907, Ontario Hydro existed, at least on paper. The battle was on to see if enough municipalities would vote on New Year's Day to contract with Hydro to make it a reality. It was the great social issue of the day, pursued with a passion that would be characteristic of energy issues later in this century. Beck stumped the countryside seeking support, talking to everyone who'd listen, gathering converts among the independent journals, building up an impressive lobbying organization.

In the previous New Year's Day elections Beck had won an important first victory when Toronto and eighteen other municipalities voted to contract with Hydro. This time the private power companies were fighting back with a vengeance, putting billboards up along the highways warning travellers of the perils of public ownership, hiring canvassers to spread free enterprise door to door, enlisting the editorial support of sympathetic newspapers and, where they weren't sympathetic, buying ads to get their message across.

The syndicate also had volunteers take up their cause, organized under the banner of the Anti-Hydro Citizens Committee of Business Men. Warning that bankruptcy was certain if Hydro won, that public debt would mount, that municipal tax rates would soar, the committee argued that the very foundations of the province would be shaken.

But they were no match for Beck's broadly based volunteer organization made up of students, housewives, engineers, and even the Canadian Women's Suffrage Association. Neither could they counter Beck's favorite example of Montreal, where huge monopoly profits were being wrung out of consumers by the very same interests that were playing the part of saviors acting in Toronto's civic interest.

The syndicate and the supporters of private utilities would lose the New Year's Day vote, arresting the cancer of private utility monopolies. On October 11, 1910, the first electric power from Niagara was switched on in Kitchener, 100 miles away. Toronto and other cities soon followed. But the people of Ontario would not stay victors for long. Cancer is a disease common to both public and private organisms. It respects no ideological barriers and is as virulent in the hands of publicly motivated socialists as in those of profit-motivated capitalists. The public would soon learn that the enemy was not greed but aggrandizement, the evil not money but monopoly, the excesses due not to size but to structure.

Once Hydro's provincial monopoly in electricity was secured, plans for a Greater Hydro emerged. It was Beck's great dream for Hydro to be not only an electrical utility but also a gas utility, not only an energy utility, but a water utility, not only a super-utility serving all the traditional municipal functions but one building, controlling and operating electric streetcars and railways to carry traffic between cities and within cities. Greater Hydro would be not only a transportation utility but a communications utility, operating the telegraph and the telephone systems. Beck proposed all this at a meeting of 700 municipal delegates just four years after Niagara became a working symbol of grandeur.

Arguing that it was absurd for the various utilities, including 450-odd telephone companies in Ontario, to be providing service, Beck reasoned that "If all these systems were joined under central control, say of the Hydro-Electric, one staff could serve all purposes."

Then he returned to his favorite theme: "Nothing is too big for us. Nothing is too expensive to imagine. Nothing is visionary with regard to this wonderful energy, this great project which is replacing the black coal of the United States. We have a great debt of $20 million but I can see that . . . the whole investment will be returned to the Government of Ontario in 15 years."

Twenty years later, Hydro owed the government seven to eight times as much as in 1914, not because the development at Niagara was uncalled for but due to the uncontrolled later expansion of the system.

Beck still commanded great public respect—at his behest 1,500 Hydro supporters invaded the provincial legislature in 1915 to present a resolution asking the government for a subsidy of $3,500 per mile for his municipal Hydro railroad scheme—but his institution's megalomania became increasingly offensive as its size, power, and ambitions encroached on others.

Franchises for private electric railways were opposed on the grounds that in electric railroads, as in the electricity that would run them, Hydro needed a province-wide monopoly.

Despite the need for war material, Beck decided to proceed with his railway plans.

When Toronto balked at giving up control over its city streets to accommodate Hydro's trains, Beck pulled all stops in pushing through his proposal for rights-of-way along the Toronto waterfront—a necessary component in the province-wide railroad network he envisaged.

Less successful was Beck's attempt—by trying to buy the Toronto Street Railroad Company when its franchise expired in 1921—to thwart the city's plan to take over public transit. But the resentment was overwhelming.

Ontario Hydro had gone out of control, expanding on all fronts at an incredible rate, putting the province into debt, acting like a sovereign state, running roughshod over other institutions and over local communities. Yet the organization, a juggernaut oblivious to all around it, had too much momentum to be easily slowed down.

To control this juggernaut—now capable of challenging the government itself—the province mobilized two royal commissions, the imported propaganda resources of NELA from the United States, and the open opposition of Premier Drury himself. Commenting on a massive power plant Hydro had built in the Thunder Bay district that supplied an embarrassingly large amount of electricity to an underpopulated area he stated, "Someone, I must say, has been guilty of lack of foresight in this development . . . The impression gradually sweeping over the people of Ontario is that another colossal blunder has been made by the Hydro-Electric Power Commission and that the district is shouldered with a white elephant in the shape of a power plant many years ahead of its time."

Hydro's growth was stunted, but not stopped, with more plants being added and electric railways being run by Hydro into the Thirties. But Hydro's structure was left intact: a cancer in remission that would re-emerge full blown in the 1970s with plans of an ambition beyond the dreams of Adam Beck. Ontario Hydro became the leading proponent of nuclear power in the free world, with an atomic program unparalleled in intensity anywhere but in the USSR. Nuclear power plants were to ring Lake Ontario, to be built eventually at the rate of one every 5½ days. To carry out this program would take money, far more money, the provincial treasurer warned, than would be available to it in all the world's capital markets by 1983. To get the necessary capital, taxes would need to be raised and funds from social services diverted.

Faced with the recurring specter of a monster Hydro rearing its still cancerous head, another premier of Ontario, Bill Davis, borrowing from history, has thrown a royal commission and a parliamentary inquiry in its path, and replaced the chairman of Hydro with a politically skilled and trusted friend in an attempt to rein the renegade organization in.

But unlike his predecessor, the premier of 1980 is short one key ally: NELA is no longer an opponent of public utilities. In the intervening years, the utility monopolies—public and private—have recognized they have far more in common with each other than they have with their nominal masters, be they taxpayers or shareholders.

When privately owned Three Mile Island (with a corporate history going back to J. P. Morgan) developed its difficulties in 1979, Hydro emerged publicly as one of its staunchest supporters. To aid its financially crippled American friend, Hydro risked the abuse of its own provincial customers and government masters. Hydro agreed, at a time acid rain had become a provincial scandal, to burn 2 million tons of coal a year to supply the private utility with electricity at cost.

The corporate mentality of public utilities like Ontario Hydro was not a product of Adam Beck; the original mentality continues in Ontario to this day. It is not limited to Ontario. In Quebec, a recent minister of energy has concluded that there are two governments—the elected government and Hydro-Québec—and of the two, Hydro-Québec is the more powerful. The corporate mentality is not peculiar to Canada. The Tennessee Valley Authority in the U.S. is rarely applauded for its sensitivity. Even before the formation of the TVA, Giant Power, a state-owned utility scheme, was being propounded by prominent conservationists and the governor of Pennsylvania as the answer to the overwhelming control exerted over the American way of life by the energy barons of the 1920s. Governor Pinchot's role was no different than Adam Beck's, and the future he saw for Giant Power was a reincarnation of Beck's vision for Hydro.

Popular magazines, technical journals, newspapers, and the broadcast media all agreed then that the rapid growth of electrical light and power was a dominant element in society, and the financiers, engineers, and technocrats behind it were the agents of the change. The conservationist reform movement, left over from Teddy Roosevelt's trust-busting era, recognized the implications of leaving control of energy in the hands of monopolies. Spearheaded by Governor Pinchot, the leading conservationist of the day, the movement organized its alternative.

"Giant Power is a plan to bring cheaper and better electric service to all those who have it now, and to bring good and cheap electric service to those who are still without it," Pinchot explained. "It is a plan by which most of the drudgery of human life can be taken from shoulders of men and women who toil, and replaced by the power of electricity."

While electric power was capable of showering upon the people "gifts of unimaginable beauty and worth," Pinchot cautioned "it is as though an enchanted evil spider were hastening to spread his web over the whole of the United States and to control and live upon the life of our people." With Giant Power "so shall we and our descendants be free men, masters of our own destinies and our own souls . . ."

Without Giant Power "we shall be the helpless servants of the most widespread, far-reaching, and penetrating monopoly ever known. Either we must control electric power or its masters and owners will control us."

Pinchot's arch Pennsylvania foe, J. P. Morgan, and the electrical interests behind him, also saw the issue as control of political power: "The very clear purpose of this plan is to take from any electric service system the benefits it has thus far accrued by reason of able management, successful financing, painstaking research work [and] courage in the replacement of apparatus," said the head of the Pennsylvania Electric Association, representing 90 percent of the state's utilities. "Private initiative is to be driven out of the electric service companies and be supplanted by a political plan based upon a socialistic theory and offering all the possibilities of the construction of a statewide, all powerful, political machine."

In fact, all that was being proposed was a change from private to public monopoly, not a change in the centralized nature of the control over electricity. What made the issue occur at that time was the recent ability to transmit electricity over greater distances than ever before, allowing control to be held, more than ever, at a central point.

Utility systems could now join transmission lines for exchanges of power and be interconnected with regional electricity networks or form new networks. Bankers saw this development as an opportunity to extend their control of holding companies; state politicians saw it as a threat to their power, fearing that—as electricity would increasingly cross state lines—the federal government would begin to assume control over the flow of electricity, in the same way it was regulating the railways and waterways.

Giant Power was a way of preserving power in the governor's seat, and Pinchot's well-meaning plan would have made any monopolist proud. To anyone burdened with a problem of herculean proportions, the temptation is great to find one solution, then apply it everywhere to answer all needs. The ability to transmit power great distances meant that Pinchot, rather than Morgan, could control it. The state of Pennsylvania, rather than Morgan's holding companies, could enter into compacts that crossed state boundaries.

Pinchot called for the establishment of giant, mine-mouth coal-burning power plants feeding a giant network of high-voltage transmission lines to contain pollution. These would replace the small plants generally located on large rivers or on tidewater. When utility engineers objected, on the practical environmental grounds that not enough cooling water existed in the coal mining regions to support the scheme (400 tons of cool water was needed per ton of coal used in producing steam for the turbines), Pinchot called for the building of dams and the construction of giant cooling towers (of the kind nuclear reactors would use decades later).

Industry again objected on practical environmental grounds. By so stretching the state's resources the economy of Pennsylvania would become vulnerable to a shortage of cooling water—the dams could not compensate for droughts and the technology for giant towers was not then available. The notion of having all of eastern Pennsylvania dependent on transmission lines extending 300 miles was also challenged—reliability of service meant more to many consumers than lower rates.

The campaign for Giant Power lost, but barely. Then, after preventing the giant public utility from adopting plans so dubious to the state on environmental grounds and economic grounds, the private utilities proceeded to adopt modified versions. Plants in Pennsylvania, as elsewhere, became larger than Pinchot envisioned, transmission distances became greater, and the massive cooling towers—transplanted to Three Mile Island—have become one of Pennsylvania's landmarks.

Co-operatives have likewise been co-opted by the utilities' love for large scale. When farmers in the U.S. were unable to convince private utilities to give them service, a populist movement emerged to set up non-profit rural electric co-operatives across the country. Co-ops were the tool for standing up to big power, a means of obtaining political and economic clout for those denied society's dividends. That was the Thirties and Forties, when electric co-operatives were struggling to get off the ground, and dependent for much of their electricity on the same private utilities that opposed their creation. To gain independence from the large private utilities, local co-ops supported the creation of their own huge, bulk-power co-ops. These large creations soon found they could relate best to other large concerns, and sought joint-partnership with the private utilities. Co-operation on large projects led to co-optation by the needs of large projects. Co-ops—now providing power to over 25 million people in areas totalling 75 percent of the American land mass—have rejected their traditional constituencies of citizens and environmental groups to publicly align themselves against clean-air and strip-mining laws (which would hamper their coal-burning goals) and privately oppose legislation as diverse as endangered species and historic preservation acts. Solar advocates who oppose nuclear generation are accused of "want[ing] to resort to primitive alternatives," and exhaustive studies of a co-op's methods of operation, if critical, are labeled "patently naive, irresponsible and biased."

In place of its old constituency a new friend has been found. Reported the nuclear industry journal, *Nucleonics Week*, "Private and public utilities traditionally regarded each other with suspicion and antagonism. Nuclear power is bringing both sides together." The goals of public and private monopolies match so perfectly, so completely overshadowing all non-monopolistic goals, that public monopolies not only behave as private monopolies but actively further the interest of private monopolies, allying themselves against individual members of the public who, by definition, form the only adversary of a monopoly.

To protect Babcock & Wilcox when its equipment for an Ontario Hydro nuclear plant was defective, Hydro agreed to pick up the $30 million price tag for repairs. To protect it from competition, contracts are awarded to Babcock & Wilcox regardless of how high their bid, often with no bidding at all. To protect Denison Mines and Rio-Algom Mines from the vagaries of the marketplace, Ontario

Hydro agreed to the purchase of $7 billion in uranium—the largest single transaction in history—on a guaranteed-profit basis. While the crown corporations like Ontario Hydro and Atomic Energy of Canada Ltd. that have a monopoly over the nuclear industry in Canada have lost billions of dollars for Canadian taxpayers, the private monopolies that supply the equipment for the public monopolies have reaped windfall profits. In effect, monopoly profits have been extracted from Canadian consumers by the Canadian crown corporations for the benefit of foreign-owned corporate monopolies.

In the course of securing these profits, crown corporations—like the private energy corporations—have become criminal elements in our society, engaging in multimillion dollar bribes and payoffs to foreign agents, such as Atomic Energy of Canada's bribes in Argentina and South Korea. They have formed secret cartels which require members to break their national laws. They withhold from the public crucial studies on safety under the guise of national security or commercial privilege. They knowingly publish erroneous information and then engage in a conspiracy of silence as the reports are allowed to be reprinted.

Dissenting views have been ruthlessly crushed and unofficial censorship imposed. A New Brunswick teacher whose Grade Four students asked an AECL public relations man about radioactive wastes was threatened with the loss of his job and harassed by mail accusing him of being "despicable." An Ottawa medical doctor who reported to the Minister of Health and the press that radioactive dust improperly dumped on a vacant lot was blowing across city streets, presenting a serious health hazard, was informed by the President of the Atomic Energy Control Board that a protest was being sent to the Ontario College of Physicians and Surgeons. The letter to the College (which regulates the medical profession) suggested disciplinary action be taken for the "unjustified and scurrilous allegations by a person whose professional standing prompts the public to accept as factual such statements." A graduate student at the University of British Columbia, who distributed a scholarly "open letter to physicians" suggesting the nuclear issue was not the open-and-shut case being espoused at a campus event, found himself physically threatened by a senior physics professor and reported to the university president. People who wrote critical letters to the editor received detailed personal letters typed on the official-looking stationery of crown corporations, taking them to task for having the temerity to voice their opinions.

These attempts at suppression were all clear instances of intimidation by government bodies: they often used extraordinarily strong language and they generally attacked the recipient's professional status.

The unofficial censorship became official when the federal government passed an Order in Council making it a crime—punishable by five years in jail or a $10,000 fine—to publicly discuss the Canadian government's uranium cartel. To make its point, the government charged a newspaper columnist (in a concurrent but different case) with exposing official secrets, although some of the information had been previously publicized in the House of Commons and on national television

No dissenting views were allowed within the industry or without. When a nuclear operator leaked information to a member of the provincial parliament showing that a Three-Mile-Island-type accident could happen at the plant he ran, Hydro branded him *persona non grata* and conducted an immediate investigation to uncover his identity. The schools were rid of critical works on nuclear power and—to this day—high school curricula do not include information critical of nuclear power on the grounds that political matters do not belong in the school system. High school teachers, meanwhile, are flown to Toronto to visit the Pickering nuclear plant and nuclear industry spokesmen hold special briefing sessions for them on a systematic basis through the teachers' "professional development" days.

University departments, meanwhile, became dependent for funding; nuclear projects and university professors became the most ardent of nuclear proponents.

The ministers in charge of energy at both the provincial and federal levels are told what to say by the nuclear industry's public relations departments, which also write their speeches, and often their correspondence as well.

The National Film Board, a crown corporation, produced several pro-nuclear films with the aid of the nuclear industry. When it decided to offer some balance by producing a film critical of the industry, it was accused of being a propaganda tool and a massive behind-the-scenes lobbying effort was undertaken to have the anti-nuclear film withdrawn.

All the while, industry spokesmen in Canada and the U.S. continue their claims of a communist conspiracy, unfazed by the fact that—at the same time Edward Teller warns that if the anti-nuclear movement is triumphant "America is doomed and freedom is doomed"—the *Wall Street Journal* reports that the Soviet Academy of Sciences considers the anti-nuclear movement a "capitalist plot."

The control of the electricity lobby is today near complete. With oil running out, in the 1970s the oil companies began shifting their massive resources into uranium and coal—the stuff of which electricity is made.

In the United States, eleven oil and gas companies already control over half of the known uranium reserves, and fourteen of the twenty major corporations now holding coal reserves are oil companies. More significantly, the oil companies also control between 35 percent and 40 percent of copper production. Copper is a key component of solar collectors, leading two U.S. congressmen to warn that the oil companies were in a position to "inhibit the marketing of solar heating and cooling systems in violation of federal antitrust statutes."

The multinationals are busy buying up solar patents and buying out small pioneering companies. Five of the world's largest oil companies are now actively involved in solar technology and research together with electronic communications companies like IBM, RCA, Texas Instruments and Phillips. Westinghouse, General Electric, and Boeing have lobbied the U.S. Congress heavily to finance a multi-billion dollar solar power satellite to convert sunlight to energy in space and relay it to earth in the form of microwave radiation. Twenty-five giant corporations receive 90 percent of the solar contracts sponsored by the U.S. National Science Foundation.

Electrical companies and oil companies—among the largest industries on the face of the earth—are merging to become one, with only the resources of individuals and small entrepreneurs in the solar energy and conservation fields left to oppose them.

The situation was not always so bleak. At one time the electrical interests were balanced by an equally vicious oil lobby, represented, more than anything else, by Rockefeller and his Standard Oil Company.

Rockefeller's contribution was not limited to his promotion of petroleum, however. In corporate history, Rockefeller is the Renaissance Man, the one credited with expanding the limits of corporate thought, creating concepts the courts had never before confronted, becoming the very model of the monopolist.

Monopolies are anathema to a free market society. Our efforts toward their control have always focused on restraining their formation. In the process we overlooked corporations—the entities of which monopolies are made. Corporations measure their success by the degree to which they've monopolized their market, yet the corporate laws we've devised are structured to encourage monopolies. Regulating the monopoly only treats the symptom of this flaw in our economy—untreated, the disease manifests itself in another manner.

The cure for monopoly lies in its corporate cause. To know how to undo the monopoly structure, we need only look at the recipe Rockefeller used to sour the corporate form for society. It is sometimes possible to tell the cause of indigestion or food poisoning after the fact, but to eliminate the guesswork it would be necessary to examine the quantity and quality of each ingredient used in the meal and to follow each step in its creation to see exactly what went wrong in the process.

# 6

## *Monopoly Money*

THE NOBILITY AND selflessness of the act were enough to make grown men cry.

"The procedure was without precedent," John D. Rockefeller explained. "We find here the strongest and most prosperous concern in the business . . . turning to its less fortunate competition . . . and saying to them. 'We will stand in for the risk and hazards of the refining business . . . Come with us, and we will do you good. We will undertake to save you from the wrecks of the refining business.'"

The modest, church-going owner of the most powerful corporation of the day was so inspired he elevated his company's activities to works of Christian charity. "The Standard was an angel of mercy," Rockefeller elaborated, "reaching down from the sky and saying 'Get into the ark. Put in your junk. We will take the risks.'"

Inexplicably, the recipients of the Standard Oil Company's missionary work were less than grateful to their protectors for the opportunity of converting to a Rockefeller-controlled organization called the South Improvement Company. "There was a pressure brought to bear upon my mind, and upon almost all citizens of Cleveland engaged in the oil business, to the effect that unless we went into the South Improvement Company we were virtually killed as refiners," said Alexander of Alexander, Scofield and Company, giving reason for the sale of his large refinery to Standard Company interests. "We sold at a sacrifice and we were obliged to. There was only one buyer in the market, and we had to sell on their terms or be crushed out, as it was represented to us. It was stated that they had a contract with railroads by which they could run us into the ground if they pleased. After learning what the arrangements were I felt as if, rather than fight such a monopoly, I would withdraw from the business, even at a sacrifice. I think we received about forty or forty-five cents on the dollar on the valuation we placed upon our refinery."

The innocuous-sounding South Improvement Company became a national scandal in 1871 when its activities accidentally leaked out after only two months of operation. Panic swept through the oil producing regions of America's Northeast, all-night meetings desperately sought ways to stop the scheme, torchlight parades, angry petitions (one was ninety-three feet long) to legislators, and threatening telegrams to railroad presidents expressed the frenzy of rage over the South Improvement Company. Overnight the name of Rockefeller became identified with infamy. A committee of the U.S. Congress would brand it the "most gigantic and daring conspiracy a free country had ever seen." For the next fifty years, Rockefeller would be the most hated man in America, his name a synonym for grasping greed.

What outraged the public was not so much the methods of the cartel that had been formed (similar methods had been used before and would become the stock and trade of future monopoly builders) but the premeditated nature of the plot, and the cold conspiracy of the strong to mow down the weak. The art of the monopoly was being formalized, its principles taken to their logical conclusion, and the conclusion was not one that appealed to a people with a sense of democracy and fair play.

But Rockefeller, as he himself was the first to admit, was being badly misjudged; his prime crime was that he was better skilled at the corporate board game than his competitors. The rules were made to be followed, he reasoned. Who was he to put himself above the laws passed by democratically elected legislatures and not take advantage of every benefit they conferred?

Were John D. Rockefeller still alive he would no doubt feel vindicated. The corporate tactics practiced then can be found in the corporate catalog of techniques currently in use. Secret rebates form the backbone of the supermarket business, and a manufacturer who does not need computer programs to keep separate his clients' rebate schemes is backward indeed. As an astute businessman, Rockefeller had used the rebate system even before the controversy over the South Improvement Company. The rebate is a corporate move that is far from complex, and pleasing to both its giver and the taker.

When Rockefeller's firm approached the Lake Shore Railroad with a guarantee to ship sixty carloads of oil per day in exchange for large rebates on freight charges, the railroad agreed out of its own self-interest. Railroads preferred a steady, high-level movement of oil to the fluctuating supply that came out of the oil regions, still highly competitive with rushes of overproduction one day and shortages the next. Rockefeller's offer meant fewer costs in juggling their locomotives, cars, and yard space. The rebates were to be kept secret, but other Cleveland refiners heard of them and protested.

The Lake Shore was forced to offer the same rebate to any other company able to give them Rockefeller's freight guarantee. Good planning by Rockefeller made this offer academic—using another tool of the monopolist's trade, he had already cornered every available tank car and container for the shipment of oil to make sure his competitors would have no means of shipping much of their cargo, let alone enough to claim the rebate.

"Rebate" became an instantly despised word in the oil regions. The oil men didn't need a translation from the French *(rabattre)* to know its meaning was "to beat down." But hated as "rebate" was, it would be a Rockefeller variation of the rebate, the "drawback," which would make the oil men livid, the South Improvement Company infamous and Rockefeller a household name.

The scheme was simplicity itself: competition would be eliminated for the mutual benefit of all participants. The railroads would form a cartel with the largest refiners in each major refining area regulating the flow of oil. Although freight rates would rise, rebates to each participating refiner would more than compensate for the inflated price. Those who refused to enter the scheme would be put out of business, unable to compete not only because cartel members were paying lower freight rates but also because those same refineries were receiving "drawbacks" from the inflated prices paid by non-members. Over $1 per barrel of the $2.56 regular rate went back to the South Improvement Company.

Secrecy, once a business courtesy, became a business necessity, formalized and ritualized: Rockefeller would frequently travel to

New York to hold clandestine meetings with other refiners and railroad tycoons like Vanderbilt, Gould, and Scott. Conducted in absolute secrecy, these meetings decided which refiners would be allowed into the scheme, providing they first took the organization's oath of secrecy:

> I, ______________, do solemnly promise upon my faith and honor as a gentleman that I will keep all transactions I may have with the corporation known as the South Improvement Company; that should I fail to keep any bargains with the said company, all the preliminary conversations shall be kept strictly private . . .

Refusal to sign this oath, Rockefeller systematically would explain to each of his competitors, resulted in further particulars of the plot being withheld, and being put out of business. Refusal to accept the South Improvement Company's terms after the plan was spelled out also resulted in being put out of business. There were only two alternatives, he told them. Collapse their businesses into his, in return for Standard stock, or try to compete and face bankruptcy. As a hedge in case any refiners decided to compete, Rockefeller assured them a hard time finding financing for their uphill battle—the executives of Cleveland's leading banks had been made stockholders in the Standard.

Once the secret of the South Improvement Company got out reaction was swift and effective. The public was outraged. The remaining oil producers and refiners united in an offensive: organizing, lobbying, agitating, and threatening until they got their way. The railroads were forced to back down, and signed new contracts with the producers, equalizing shipping rates. Predictably, as might happen today under similar circumstances, a bill revoking the charter of the South Improvement Company was rushed through and signed with much fanfare.

The independents had taken on the Standard and won, they'd flexed their muscles and the monopolists backed off. The euphoria of their success soon evaporated, however, when the oil men surveyed what had been salvaged. Of Standard's twenty-five competitors in Cleveland, Rockefeller had managed to buy all but three before losing the battle to the independents. Cleveland was now under his control.

Only those who valued their principles over economic survival could have refused to join the South Improvement scheme. So strict were the rules of the game that no exceptions could be made for anybody—not even for Frank Rockefeller, John D.'s youngest brother, who refused to join, though warned by his brother that: "We have a combination with the railroads. We are going to buy out all the refiners in Cleveland. We will give everyone a chance to come in . . . Those who refuse will be crushed. If you don't sell your property to us, it will be valueless."

The naive younger brother did not understand the nature and needs of monopolies and became bitter for the rest of his life, testifying publicly against the senior Rockefeller on several occasions and eventually removing his two childrens' bodies from the family burial plot, vowing "No one of my blood will rest upon land controlled by that monster John D. Rockefeller."

Had the brothers' positions been reversed, the results would have been the same. Monopolies are not wished by people but develop their own characteristics, growing inevitably to whatever extent their corporate genes have been pre-programmed.

Ironically, Rockefeller did not dream up the South Improvement Company, or most of the other schemes that would make him the world's first billionaire. Rockefeller subscribed to Andrew Carnegie's theory that "pioneering don't pay." Rockefeller would always leave the innovating to others.

He wasn't even an oil pioneer. He entered the field as a silent partner in a refinery only after petroleum production had run up into the millions of barrels and only on the understanding that oil would be a secondary interest. Rockefeller considered oil a non-renewable resource, and so an unstable commodity on which to base his business.

Rockefeller's success was due less to outcompeting his rivals than to tying their corporate hands to ensure they couldn't compete at all. His genius lay not in the production, sales, or marketing of oil but in understanding the structure of corporations. But it was not by chance that oil made his success. Oil was the most monopolizable of resources.

In retrospect, his formation of a great monopoly would seem the result of an unbelievably diabolical plan. Yet Rockefeller had no preordained plot. He merely recognized opportunities presented to him, as hundreds of others were recognizing opportunities in less monopolizable commodities. Rockefeller and the Standard Trust he formed were no more than a part of the great monopoly juggernaut that was in the process of being set up by seemingly sensible legislation. "None of us ever dreamed of the magnitude of what became the later expansion," Rockefeller admitted.

At the start of Rockefeller's next campaign, the Standard still competed with twenty-seven refiners in the oil regions, twenty-two in Pittsburgh, fifteen in New York and twelve in Philadelphia. By 1877 the Standard had no competition in the regions, Pittsburgh, or Philadelphia, and only a handful of independent refiners were left in New York. By 1878 Standard was producing 33 million of the 36 million barrels of oil per year refined in the United States. By 1880 the Rockefeller monopoly refined 95 percent of America's oil.

It wasn't enough.

After Standard had secured its control over the refining and transporting of oil in the U.S. Northeast it went after the marketing end nationwide, ruthlessly exterminating small independents and local wholesaling agents. Parceling the country into sales territories, Rockefeller sent Standard affiliates into small towns and villages to undercut those who had been selling kerosene before them.

Anyone in the monopoly's path was crushed, including one who thought he could take on Standard at its own game.

George Rice, an Ohio refiner who'd been marketing small amounts of oil in the South for many years, decided to fight back by undercutting Standard, hoping to wait his opponent out. But even well-established customers stopped ordering from Rice—Standard had spread the word that up to $10,000 would be spent to break anyone who bought Rice's oil. This secondary boycott, despised when practiced by unions today, worked with ruthless efficiency. A Louisiana retailer named Wilkerson and Company defied the Rockefeller dictum and soon felt Standard's wrath. The railroad that delivered Wilderson's oil was contacted: "Wilkerson and Company received car of oil Monday 13th—70 barrels which we suspect slipped through at the usual fifth class rate—in fact we know it did—paying only $41.50 freight. Charge $57.40. Please turn another screw."

Misunderstanding Standard's mandate, one rival labeled Rockefeller "the Mephistopheles of the Cleveland Company" but Rockefeller rightly refused to take credit for his accomplishments.

"They could not hope to compete with us," he eulogized after the demise of his competitors. "We left them to the mercy of time." The evolution of the corporate cosmos, Rockefeller felt, followed laws as inexorable as the laws of nature. The lawbreakers were not the Standard group but those who refused to join it and benefit from the new reality.

The South Improvement Company could not have been possible without new legislation in Pennsylvania which permitted the chartering of a new kind of company, a holding company, which allowed its owners to control stock in companies both in and outside the state and so escape surveillance by state regulators. The charter, broad and vaguely defined, permitted the company's owners to conduct any business any way they chose.

But even without the holding company, much could be done by working around existing legislation. To expand beyond his Cleveland confines, Rockefeller enlisted two allies, one refiner in the regions and one in New York, to buy up refineries.

The Standard could not buy the companies directly because of state laws that quaintly maintained a corporation couldn't exist in two states at once. But, by financing existing refineries to secretly buy out the local competition, no one knew what had happened until it was too late. An immaculately conceived operation, it was self-financing. Rockefeller's allies received Standard stock for their efforts; illicit rebate money from the competition itself was used to buy them out.

The Standard Oil Company's expansion began to bump into the law of diminishing returns as less and less of the graspable market remained to be absorbed. The corporate structure, so necessary to Standard's past growth, was now restraining Standard's fortunes. Rockefeller recognized that the current concept of the corporation had become obsolete and conceived a new one—the trust. Though known as the Standard Trust, the new organization had no official name or charter . . . it was merely the common law concept of one party holding property for another. But it enabled the monopoly builders of the era to cross state lines and slickly penetrate virgin territory.

Before the creation of the Standard Trust in 1882, Rockefeller's expansion was menaced because Standard could not own and operate manufacturing facilities outside the state. Under the trust, all outstanding Standard stock was transferred to nine trustees, who were then empowered to dissolve participating organizations and set up new ones in each state—Standard of New York, Standard of New Jersey and so on, forty corporations in all, of which fourteen were wholly owned. Its intricate workings and elaborate complexities, guarded by a maze of legal structures, made it inpenetrable to public investigation and exposure. As a noted historian of the day recognized, "you could argue its existence from its effects, but you could never prove it."

Only once did it appear the Standard might be thwarted. Attempting to break Rockefeller's stranglehold on the railroads and oil distribution system, the Tidewater Company undertook a daring gamble—it decided to build a 110-mile pipeline from the oil regions to the sea, an engineering feat likened at the time to the building of the Brooklyn Bridge. Unawed, Rockefeller threw every impediment in the way, buying up rights of way to block its path, intimidating workmen, even sabotaging the pipe itself.

But the Tidewater found its way to the sea, with its own pipeline to markets Standard couldn't touch. It had withstood the challenge from without, only to be destroyed from within. A Rockefeller aide bribed his way into the company and ground it to a halt, besieging it with stockholder fights and other problems until, finally, it too sold out to the Standard. Tidewater's own corporate structure led to its undoing, and again the Standard would be spat at in disgust.

The public, not understanding as Rockefeller did that "the Standard Oil trust is a beneficial organization to the public," viewed it with intense hostility, eventually forcing government to act first with investigations and later with legislation. A New York senate committee in 1888 called the Standard Trust "the type of a system which has spread like a cancer throughout the commercial system of the country," and the New York *World*, after Rockefeller's testimony before the senate committee elaborated upon his definition of a trust: "A philanthropic institution created by the benevolent absorption of competitors to save them from ruin, combined with the humane conservation and ingenious utilization of natural resources for the benefit of the people."

As the Standard was consolidating its grip over its American monopoly, its monopoly overseas was being threatened by Russian imperialism. For a quarter of a century after the Titusville gusher, the U.S. had been the only exporter of oil in the world (with Standard accounting for 90 percent of that exportable surplus). Standard had fought ferociously for its overseas monopoly, extending to foreigners its corporate culture of price-fixing, bribery, and buy-outs, defying governments abroad as routinely as state legislatures at home.

This was all threatened by the opening of the great Baku fields on Russia's Caspian coast in 1883.

That year Standard got together with the Russian oil czars, the Nobel brothers and the Rothschild family—the leading world monopolists—to work out a scheme between them for parceling out the whole of the refined markets of the world. But the precommunist Russian government refused to go along with the Standard plan, which would have forced all Russian refiners to join in one organization for the purpose of selling to the export market. The Russians, Nobels, and Rothschilds took on Standard. By 1888 Russia had overtaken America in the production of crude, had cornered 30 percent of the English market for kerosene, and was threatening the balance of Standard's international market in Europe.

Standard fought back, cutting out the European importing firms it had previously used for a system of foreign affiliates—the Anglo-American Oil Company in England, the Deutsch Amerikanische Gesellschaft in Germany, and others. When more negotiations with the Rothschilds failed to "rationalize" the international oil market, price-cutting and secret purchases of stock were used. The Rothschilds were Rockefeller's match, however. Though exports of U.S. oil to Europe grew five-and-a-half times from 1884 to 1889, Standard would maintain only 60 percent of the market until the Great War of 1914 changed the ground rules altogether.

Still, Standard pressed on in Canada, in the Middle East, and in Southeast Asia, usually aided by "one of our greatest helpers . . . the State Department in Washington." Rockefeller gratefully acknowledged that "our ambassadors and ministers and consuls have aided to push our way into markets to the utmost corners of the world."

With an inconsistency that's a characteristic of bureaucracies, the sterling co-operation Standard received from the U.S. government abroad (where Standard was virtually allowed to write its own foreign policy) was never matched at home.

By the turn of the century the trust was besieged by one investigation after another, constantly defending itself against charges that the various trustees were acting in concert, in violation of the Sherman Anti-Trust Act of 1890. It was little consolation that the trustees could perjure themselves freely, knowing the impossibility of disentangling the structural maze of secrecy and legal deception Standard's lawyers had devised. The tedium of continually answering "no sir, we had no better [railroad] rates than our neighbors" or of suffering from repeated bouts of amnesia ("I could not say" and "I cannot recollect about that") was exacerbated by newspaper accounts advising "The art of forgetting is possessed by Mr. Rockefeller in its highest degree." True consolation would not come until 1911, when the Standard Oil trust was broken up into thirty-nine different companies by Teddy Roosevelt's much vaunted trust busters.

Within a week after the trustees of the old Standard Trust had received their shares in the new Standard companies, these companies were traded on Wall Street for the first time. The result was the biggest bull market the stock exchange had ever seen for a single stock. Standard of New Jersey jumped from $260 to $580 a share; Standard of Indiana from $3,500 to $9,500. In the next five months, $200 million was added to the value of Standard stock. The same men who owned the old trust owned the new companies, and they would run them as if nothing had changed. The only difference seemed to be that the owners were now a little bit richer, and the corporations they controlled a lot more powerful. A blistering 1500-page U.S. Senate report a dozen years after the breakup deplored the "complete control by the Standard companies . . . in some respect the industry as a whole, as well as the public, are more completely at the mercy of the Standard Oil interests now than they were when the decree of dissolution was entered in 1911." Said Teddy Roosevelt just one year after the breakup: "no wonder that Wall Street's prayer now is—'Oh Merciful Providence, give us another dissolution.'"

The corporate structure had emerged triumphant despite the success in wresting it from individual control—not surprisingly, since it had been the corporate structure all along that had made

Standard a success, not Rockefeller's personal genius in petroleum.

When Standard stopped Tidewater's underground attempt to beat Rockefeller's monopoly above ground its technique was internal infiltration and sabotage—corporate guerilla warfare on the field had failed. True competition—matching innovation with innovation or price reduction with price reduction—was never a factor. Only after Tidewater's pioneering pipeline enterprise was taken over would Standard steal its advantages and make them its own, revolutionizing the industry.

Throughout Standard's climb, the technique of secret rebates and secret deals was used to override the superior position of rivals. With Standard everything was clandestine, dealt with as state secrets. Rockefeller would often tell associates "you'd better not know" or "if you don't know anything, you can't tell anything." Spies were recruited in rival organizations to make daily transcripts of activities. Where shipments went and what they cost to produce were reported so thoroughly that one competitor complained "before I'm able to ship these the Standard has gone there and compelled those people to countermand those orders . . . If they don't, Standard will put the price of oil down to such a price they won't be able to afford to handle my goods."

Codes were used in telegrams and letters during the monopolization of the Northeast. For example "doxy" meant Standard, "doubters" referred to refiners, "mixer" meant freight drawbacks, and "druggist" was the city of Philadelphia. Even corruption was made easier through the corporate structure, which cloaked payoffs and bribes as business expenses.

Though the "Standard had done everything to the Pennsylvania legislature except refine it" (in the words of an eminent contemporary scholar) it was doing no more than fulfulling its mandate to make profits—its sole mandate as acknowledged years earlier by a chief justice of the Supreme Court. Two decades later, the Supreme Court judge who presided over the breaking up of the Standard Trust stated in his decision that he was "irresistibly driven to the conclusion that the very genius for commercial organization soon begot an intent and purpose to exclude others."

A new kind of corporate creature had been created, one with an independent existence, whose success depended not on innovation or quality but on the control exercised over us by a legal structure we ourselves had created.

How durable this corporate being became can be seen by its longevity—most of the great corporations of today (General Motors, Canadian Pacific, INCO, Union Carbide, Bell, General Foods, General Electric, United States Steel, Ford, RCA, DuPont, Armour, International Harvester, Westinghouse, and on and on and on) were products of those times, effectively stifling new business ventures for a half-century and more. How devastating monopoly power became can be seen by the unchecked influence it currently holds in world affairs, an influence unimpeded by love of country or love of God, guided only by its social responsibility to the almighty dollar.

How devastating monopoly power became when in control of energy is a phenomenon that all but defies belief in a conditioned public desirous that it were not so.

# 7

# *Fuel Wars*

AYATOLLAH KHOMEINI AND the hostage incident . . . Petrocan . . . lineups at the gas pumps . . . hijackings . . . oil shortages and the search for alternative fuels . . . anti-nuclear demonstrations . . . a shift of our economy to the tar sands and oil shale of the West . . . the economic takeover of entire towns by Arab oil sheiks . . . soaring fuel bills . . . the drive for energy self-sufficiency . . . . the 1973 Yom Kippur War . . . the Soviet coup in Afghanistan . . . the Olympic boycott of 1980 . . . the embargo on U.S. wheat sales to Russia . . . the fall of the Clark government . . . the massacre at Mecca's Great Mosque . . . oil spills . . . Arafat at the U.N. . . . the storming of the U.S. embassy in Pakistan . . . hostility toward Iranian students . . . Since the 1973 OPEC oil crisis, not a day has gone by that the foreign policy implications of energy haven't affected us at home.

Energy affects our comforts, our lifestyles, our economic well-being, our national pride, our personal independence, our collective independence, our choice of friends, our politics. This is no coincidence. A boycott of any other commodity could not have affected us so profoundly—only energy is indispensable to everything. This is also nothing new. Since the discovery of oil—the most portable of fuels—energy has always been a pivotal point around both domestic and foreign policy, an ultimate concern of all political leaders, because it controls the ultimate determinant of politics—the ability to wage war.

If this concept seemed abstract to governments at one time, its significance was not lost on anyone after World War I. "He who owns oil, will own the world," warned France's wartime oil commissioner. "The Allies," declared Lord Curzon in his famous postwar statement, "had floated to victory on a wave of oil." U.S. President Coolidge, stressing the strategic importance of world oil supplies, created the Federal Oil Conservation Board and placed on it the secretaries of war, navy, interior, and commerce. Formally identifying oil with its national interest, the U.S. also helped establish a syndicate of Standard and the other major U.S. oil companies to control foreign oil reserves worldwide.

Though political leaders used the multinational energy monopolies in the national interest, the corporate interest of the energy monopolies did not always correspond to the national interest of their home countries.

By being fully responsible to no one country, the multinationals, in many respects, saw themselves as more stable and above the trivial considerations of politics and politicians, which change with each change in government. As supranationals—sovereign economic states—they follow their own policies and acknowledge loyalty to no particular country. Henry Ford II, chairman of Ford Motor Company, bluntly stated, "We don't think of ourselves as a national company anymore. We are definitely a multinational organization . . ." Putting things into perspective for a group of worried stockholders on the eve of Germany's invasion of Poland in 1939, GM's chairman Alfred P. Sloan, Jr., explained that his corporation was "too big" to be affected by "petty international squabbles."

The German chemical multinational, I.G. Farben (a monopoly spearheaded by such household names as Bayer and BASF) became more Nazi than the Nazis themselves, methodically planning the takeover of the chemical companies of Czechoslovakia, Austria, Poland, France, Britain, and even the U.S., in step with Hitler's planned takeover of their governments. It was I.G. that chose to build a great industrial complex at Auschwitz for the production of synthetic oil and rubber. When the slave labor of concentration camp inmates provided by the Nazis proved inadequate for I.G.'s needs, the company set up and ran its own concentration camp there. The Auschwitz plant was the boldest investment the company had ever made—close to a billion marks of its own money. It refused to jeopardize this investment by leaving its labor supply in government hands. So massive was I.G.'s installation that it used as much electricity as did the entire city of Berlin. Twenty-five thousand camp inmates died constructing it.

At the Nuremberg Trials, where the directors of I.G. were prosecuted, the setting more resembled an anti-trust suit than a trial for slavery and mass murder. The twenty-three defendants present were among the industrial elite of Germany, not Hitler's black-and-brown-shirted bullies. Representing that extraordinary combination of commercial capability and scientific genius unique to a private industrial enterprise, these were the same directors who served on the boards of the most prestigious corporations in Germany and abroad, who had made I.G. pre-eminent in the world of technology and commerce, and who were treated everywhere with awe and admiration. Like their counterparts elsewhere, they were leading supporters of culture, religion, and charity.

The indictment from Nuremberg reads in part: "All of the defendants, acting through the instrumentality of I.G. . . . participated in . . . the enslavement of concentration camp inmates . . . the use of prisoners of war in war operations . . . and the mistreatment, terrorization, torture, and murder of enslaved persons. In the course of these activities, millions of persons were uprooted from their homes, deported, enslaved, ill-treated, terrorized, tortured, and murdered."

General Telford Taylor, chief prosecutor at Nuremberg, accused the I.G. employees "of major responsibility for visiting upon mankind the most searing and catastrophic war in human history . . . of wholesale enslavement, plunder and murder" and, in his opening statement, correctly anticipated their defense. "The defendants will, no doubt, tell us that they were merely overzealous, and possibly misguided patriots. We will hear it said that all they planned to do was what any patriotic businessman would have done under similar circumstances . . ."

An excess of patriotism was not an explanation to be easily believed. Before the war, I.G. resisted many of the Nazi's policies that ran counter to its corporate goals, so strongly that Hitler considered the company all but treasonous. Had he needed I.G. less, he might have done without them.

I.G. was really no different than any other multinational earnestly furthering its own sovereign goals. The board's first loyalty was to company, not country. Luckily for I.G., it was able to capitalize on a foreign policy whose goals were consistent with its corporate goals. Less luckily for I.G.'s multinational partners, their corporate goals were at odds with the foreign policy of their home countries. I.G.'s corporate partners included Standard Oil and General Motors.

The importance of I.G. to Germany can not be overestimated. With virtually no oil wells of its own, Germany had been strangled by the British fleet during World War I. I.G. developed the technology to produce synthetic oil from coal, leading Gustav Stresemann, chancellor and foreign minister during the Weimar Republic, to say, "Without I.G. and coal, I can have no foreign policy." Hitler repeated that sentiment to I.G. officials some years later, somewhat more emphatically . "Today an economy without oil is inconceivable in a Germany which wishes to remain politically independent. Therefore German motor fuel must become a reality, even if this entails sacrifice. Therefore it is urgently necessary that the hydrogenation of coal be continued."

But I.G. didn't have all the expertise it needed on its own, and in the mid-Thirties, at the urgent request of Nazi officials, it joined with the Ethyl Corporation (owned 50 percent by Standard, 50 percent by General Motors) to form Ethyl GmbH to build and operate tetraethyl plants in Germany.

Without these plants, as records captured after Germany surrendered revealed, "the present method of warfare would be unthinkable . . . The fact that since the beginning of the war we could produce lead-tetraethyl, a vital gasoline additive, is entirely due to the circumstances that shortly before the Americans had presented us with the production plants complete with experimental knowledge."

After the war, an investigation ordered by General Eisenhower concluded I.G. was indispensable for the German war effort, confirming that without it Hitler could never have embarked on the war or come so close to victory.

None of this was news to Ethyl Corporation, which (in case it didn't appreciate the consequences of its impending merger) had been urged by DuPont to abandon its plans: "It has been claimed that Germany is secretly arming. Ethyl lead would doubtless be a valuable aid to military aeroplanes. I am writing you this to say that, in my opinion, under no condition should you or the Board of

Directors of the Ethyl Corporation disclose any secrets or 'know-how' in connection with the manufacture of tetraethyl lead plants in Germany."

Inexplicably, the advice was ignored. Before World War II, as Germany was preparing its invasion of Czechoslovakia, the Luftwaffe realized its reserve supplies of tetraethyl lead were dangerously low and asked I.G. to borrow 500 tons from Standard, its U.S. partner. "To conduct modern war," said an I.G. official, "without tetraethyl lead would have been impossible." The rush delivery from Ethyl Export Corporation, a Standard Oil affiliate, was completed before the Czech invasion.

Standard, intent on maintaining its excellent relations with I.G., refused to let the next world war get in the way. Its relationship with I.G. continued up to, during, and after World War II, in the same conscientious, even kindly, manner that it had in the past.

In 1934, attempting to improve the poor image that I.G. and the Third Reich had in the U.S., Standard Oil sent Ivy Lee, Rockefeller's personal P.R. man, to Germany. Standard's attempt at international relations was poorly received. Lee came home to face a severe inquisition by a Special House Committee on Un-American Activities, and when the committee testimony was released in July, shortly after the "night of the long knives," (Hitler's blood purge of the SA) the result was a headline which read, "Lee Exposed as Hitler Press Agent."

During this period, I.G.'s subsidiary in the U.S. was American I.G. The board of directors of this giant included Walter Teagle (president of Standard Oil), Edsel Ford (president of Ford), and Charles Mitchell (chairman of Standard's bank, the National City Bank).

Under investigation by the U.S. Securities and Exchange Commission, officials of American I.G. from 1934 to 1936 held that there was no German front for their company. The most striking spectacle was that of Teagle in testimony before the Securities and Exchange Commission in 1938, repeatedly claiming, to the obvious incredulity of the commission, that he did not know who the owners of the company were. Teagle's testimony proved to be such an embarrassment the company then change its name to General Aniline and Film Corporation (GAF), the name it is still known by today. Standard would later change its name to Esso and Exxon.

In 1938 Standard successfully dissuaded Goodyear and Dow from developing a synthetic rubber capacity of their own out of loyalty to I.G., despite their knowledge that the spread of the war to the Pacific would make America's supply of natural rubber vulnerable (as happened). After Pearl Harbor, rubber was tightly rationed in the U.S., and recycling of old rubber products became a national pastime. Internal correspondence spelled out where Standard's loyalty lay when it decided to lie to the U.S. firms: ". . . since under the agreement they [I.G.] have full control over the exploitation of this process [of making synthetic rubber], the only thing we can do is . . . loyally preserve the restrictions they have put on us."

But the greatest test of friendship would come a year later, as war was breaking out and I.G.'s holdings in the U.S. were in danger of being seized by the American government.

To protect I.G. financial interests and preserve the I.G.-Standard Oil relationship, the two companies engaged in a frantic camouflaging operation involving dummy companies, Swiss intermediaries, personal conspiracies, and falsified documents. I.G. and Standard were working against the clock, also fearful that their secrets would fall into the hands of the French or British, who would then use them in wartime without respect for I.G. and Standard's property rights.

Most, but not all, of the flurry of negotiations were completed on time. The balance, completed after war broke out, was backdated. As I.G. and Standard well knew, Americans were likely to take a narrow view of their operations.

Since early 1941, the Department of Justice had been investigating the I.G.-Standard cartel, and after Pearl Harbor the government was determined to indict the Standard Oil companies, I.G. Farben, and their principal officers for a conspiracy to restrain trade and commerce in the oil and chemical industries, including synthetic rubber and synthetic gasoline, throughout the world. Standard protested the suit, on grounds of patriotism. Their energies needed to be devoted entirely to the war effort, they said, not dissipated in some lawsuit. The government was not moved. Standard was sued, the anti-trust complaint indicating that America's then current rubber crisis could be traced, at least in part, to the I.G.-Standard cartel. Standard pleaded "no contest" (on patriotic grounds, again, they explained they would not fight it). The fines totalled $50,000.

The company's patriotism had come into question shortly before this time due to correspondence from Standard Oil's own files tabled before the Senate Special Committee to Investigate the National Defense Program. Signed by a Standard Oil vice president, the document perfunctorily explained "representatives of I.G. Farbenindustrie delivered to me assignments of some 2,000 foreign patents and we did our best to complete plans for a modus vivendi which would operate through the term of the war whether or not the U.S. came in." That last phrase sent a shudder through the committee room, and caused the committee's chairman, Harry S. Truman, to leave the hearing in a huff, snorting, "I think this approaches treason."

Despite what was clearly an anti-Standard mood in the U.S. during the course of the war, and despite the risk of a backlash, Standard had too much at stake to stand by and do nothing when the U.S. government seized the GAF corporation in April, 1942, on grounds that it was an enemy-controlled corporation. Standard sued, charging that the patents and stock had been wrongfully seized.

The emotionally charged trial, which began just two weeks after Germany's surrender, stunned Standard's directors when a surprise U.S. witness clicked his heels in salute to them. The witness was August von Knieriem, chief legal counsel of I.G., who had been captured by General Patton's Third Army. Appearing in court with the original contract (on which was written "Nachkreig Camouflage," German for "Postwar Camouflage"), von Knieriem provided convincing evidence. Another I.G. document helped seal the decision: "By this transfer . . . the patents in German possession will be removed from enemy seizure . . . We consider it right to transfer the patents to an American holder who is on friendly terms with us and who will cooperate with us on a friendly basis in the future."

GM and Ford, of course, had a responsibility to their shareholders. Refusal to co-operate would have resulted in confiscation by the Nazis and serious financial losses to the corporation. Refusal not being an option consistent with their corporate responsibility to their stockholders, GM and Ford decided to make the best of the bad situation others had created, and maximized profits by supplying both sides with the material needed to conduct the war.

With the outbreak of war in September 1939, automotive plants were fully converted to wartime production of trucks and aircraft. GM managed the conversion of its 432-acre Opel complex in

Russellsheim to warplane production in just four months. Right through to 1945, this one facility assembled 50 percent of all the propulsion systems produced for the J.U.-88 medium-range bomber, considered the Luftwaffe's most important bomber. Also produced at Russellsheim were jet engines for the M.E.-262, the world's first operational jet fighter, considered by military experts "the most important military aircraft to come out of Germany." Its top speed of 540 miles per hour was a full 100 mph faster than the American P-510 Mustang (the fastest piston-driven fighter the Allies had).

GM's contribution on the ground was even more significant. Together with Ford, they accounted for close to 90 percent of the armored "mule" three-ton half-trucks and over 70 percent of Germany's medium- and heavy-duty trucks. According to American Intelligence reports, these vehicles formed "the backbone of the German Army transportation system."

Before the war and during the war, GM was in complete management control of its German operations. As 100 percent owners of Adam Opel, GM selected the company's board of directors and appointed the management, which supervised wartime operations of all Opel plants. Alfred P. Sloan, Jr., board chairman of GM in the U.S. and GM vice presidents Mooney, Smith and Howard all served on the GM-Opel Board of Directors throughout the war, apparently without any German interference until November 25, 1942, when Professor Dr. Carl Luer was appointed as an administrator to the Russellsheim plant. But even then, as the Darmstadt Provincial Court of Appeal stressed in its appointment of Luer, "the authority of the board of directors shall not be affected by this administrative decision." The management during the war remained essentially the same as it had been before the war (with the exception of American personnel) at Russellsheim and GM's other German facilities at Berlin, Brandenburg, Aachen, Breslau, Dusseldorf and Magdenburg.

From a business point of view, however, Sloan's hands were tied. After Germany declared war on the United States on December 11, 1941, demand for automobiles was almost non-existent. Without conversion of Russellsheim to airplane production the Opel plant would have faced economic collapse. In keeping the plant going, Sloan was protecting American investors. The better to protect them, trade between the enemy countries took place: for the entire duration of the war, production efficiency was maintained as communications and material continued to flow between GM and Ford plants in Allied and Axis countries.

Corporate interests were being protected in another way as well. It was by no means certain that the Allies would win the war. Had the Nazis won, GM and Ford would have emerged as impeccably Nazi, major instruments of Hitler's achievement. In much the same way that corporations protect their political interests in peacetime by investing in both political parties before an election, by balancing their investments in wartime, GM's and Ford's position was secured regardless of the outcome. In doing so, the corporations were accountable to their stockholders, but unaccountable to the citizens of any country, in effect acting as private governments with great power to influence the course of World War II.

Nor were they embarrassed by their activities, which were executed with a straightforward business efficiency. In 1942 GM sought and got a $35 million tax write-off from the U.S. Treasury for its German operation. After the war GM and Ford demanded compensation for the wartime damage Allied bombing did to their Axis plants in Germany, Austria, Poland, Latvia and China. General Motors received $33 million, Ford less than $1 million. Their rebuilt plants at Russellsheim and Cologne have enabled GM and Ford to control two-thirds of the German automobile market while GM's truck plant in Brandenburg, East Germany, and Ford's facilities in Budapest, Hungary, are significant factors in these communist economies.

The court decided that all transfers of assets from I.G. to Standard after the 1939 agreement were a sham designed to create the false appearance of Standard ownership of property interests that both parties continued to regard as owned by I.G. "The court is satisfied," U.S. Judge Wyzanski wrote, "that the over-riding real agreement of the Jersey group [Standard] and I.G. was that . . . these transfers both of legal title and equitable interests were to be null and void at the pleasure of I.G. and the parties intended that after the close of World War II, the Jersey group and I.G. would make whatever deal then deemed to be appropriate."

This decision was upheld and strengthened in circuit court two years later when Judge Charles Clark ventured the opinion that Standard could have been considered an enemy national in view of its relationship with I.G. Farben after the U.S. and Germany had become active enemies. Yet these suits only settled some matters. Many I.G.-Standard agreements were left intact, and Standard was able to buy back from the government some of the property it had lost in court for a mere $1.2 million.

I.G. itself also fared extremely well. Despite Eisenhower's decision that I.G. must be broken as "one means of assuring world peace," I.G. was kept together and three weeks after Germany became a soverign state (May 1955) and the Allied occupation forces left, 450 stockholders of the old I.G. Farben met for the first time since the defeat of Germany. I.G. directors who had been sentenced for war crimes ten years earlier became directors of I.G.'s successor companies: Friedrich Jaehne was elected chairman of Hoechst in June 1955. Fritz ter Meer, the only war criminal to be convicted of both plunder and slavery, was elected chairman of the supervisory board of Bayer.

These companies were no small concerns. The profits of BASF, Hoechst, and Bayer already surpassed those of I.G. at its heyday, and their stock was valued at more than 15 percent of all stock listed on the West German Stock Exchange in 1956. By 1977, Hoechst had become the largest company in Germany. BASF was larger than DuPont, and Bayer rivaled DuPont in size. Hoechst, BASF, and Bayer are now among the thirty largest industrial corporations in the world.

The role of I.G. during World War II was primarily in support of the Nazi government (although in defending its conduct in giving away secrets to the Germans, Standard claimed that I.G. had also passed along secrets to Standard). The role of Standard was primarily to protect its financial interests (although it not only provided the Germans with the technology to produce their own fuel but also supplied Germany's Axis partners with fuel from its Romanian and Hungarian oil fields).

The role of General Motors, Ford, and Chrysler was less ambiguous. In the U.S., these companies were national heroes, producing war equipment at remarkable rates in support of the Allied war effort. With an unparalleled sense of fair play, these multinationals were doing the same thing for the Axis war effort.

The Big Three's involvement in Europe began in the 1920s and 1930s when they undertook an impressive program of multinational expansion. By 1929 GM owned Adam Opel, Germany's largest auto firm, and by the mid-Thirties the Big Three's subsidiaries had also spread to politically sensitive areas like Hungary, Latvia, Austria, Romania, and Japan.

GM, in particular, as owner of Germany's largest automotive plant, had a large part to play. In 1935, accepting a site which military officials advised would be less vulnerable to enemy air attack, GM's Opel subsidiary located a new heavy truck facility at Brandenburg to supply the Wehrmacht with Opel "Blitz" trucks. For this and other wartime preparations, in 1938 Hitler awarded GM's chief executive abroad the Order of the German Eagle (first class). Ford's chief executive, meanwhile, was receiving the Nazi German Eagle (first class) for its truck assembly plant in Berlin. According to U.S. Army Intelligence, the "real purpose" of the plant was to produce "troop-transport-type" vehicles for the Wehrmacht.

To their credit, the energy multinationals were acting not only out of greed but also out of principle—big business in America believed in the virtues of fascism. The 1930s was a time of social upheaval. The Depression had made desperate men of otherwise passive persons; there was talk of socialism in the air and the American people had elected as president Franklin Delano Roosevelt, a man viewed by the big business community with the deepest distrust. Contrasting the gloom of America were glowing reports of "the miracle of the twentieth century" that GM president William Knudsen saw in Nazi Germany, where he was greeted by Goering.

Earlier in the decade, G.E. president Gerard Swope had won wide acclaim in the business community (including the U.S. Chamber of Commerce, the National Association of Manufacturers, and the National Industrial Conference Board—three of the most powerful business organizations) for his 1931 blueprint for turning America into a fascist state. Later called the Swope Plan, it called for the compulsory cartelization of all major American corporations into federally controlled trade associations for each industry. Central planning would be carried out by a national economic council of corporate leaders and "responsible" union leaders. Praised as "economic Mussolini," the Swope Plan, which paralleled plans Hitler was at the time proposing, could not have fit in better with the business mood of the day. A poll of the members of the Chamber of Commerce showed 90 percent to favor centralized planning, and everyone seemed to support G.E. chairman Owen Young in his call for "strong government."

One financier wrote: "If any country in continental Europe were confronted by a crisis of the character indicated in the United States today, there would be a demand for a dictator. Unless conditions improve—and improve very soon—we are not certain that we ourselves may not be involved in a situation that may call for just such a step." Another financier, who was soon elected president of the U.S. Chamber of Commerce, asked for the right "to suspend the operation of existing laws and to provide for emergency measures" which could be administered by a business economic council. Al Smith, the political *protégé* of both DuPont and General Motors, was more direct, suggesting that "to say the least, a mild form of dictatorship, honestly operated, honestly intentioned, must be set up."

And it almost was. Discontent with F.D.R. had grown with each piece of "socialistic" and "unconstitutional" legislation he passed (requiring public disclosure of information, creating the Tennessee Valley Authority, guaranteeing collective bargaining), and the corporate leaders of the country decided to set up an alternative—an independent political organization "encouraging people to work, encouraging people to get rich, showing the fallacy of communism in its efforts to tear down our capital structure, etc."

Speaking was Jacob Rascob, one of the most powerful men in the country, a former National Chairman of the Democratic Party, and a financier with the closest of links to the Morgan, DuPont, and General Motors organizations. Rudely deposed as National Chairman upon F.D.R.'s election as president, alarmed by Roosevelt's radicalism and supported by some of the biggest corporate powers in the country, Rascob became the recruiter for the new movement. Writing to the man who had organized a spy system for DuPont during World War I, Rascob stated "I know of no one that could take the lead in trying to induce the DuPont and General Motors groups, followed by other big industries, to definitely organize to protect society from the suffering which it is bound to endure if we allow communistic elements to lead the people to believe that all businessmen are crooks, not to be trusted, and that no one should be allowed to be rich . . . The reason I say that you are in a peculiarly good position to do this is that you . . . are in a position to talk directly with a group that controls a larger share of industry through common-stock holdings than any other group in the United States."

Five months later, on August 15, 1934, the corporate alliance was consummated. The American Liberty League was founded to "combat radicalism," "uphold the Constitution" and "preserve the ownership and lawful use of property when acquired." Its officers and executive included the likes of Rascob, and Irenee DuPont (whose family controlled the DuPont and General Motors fortunes).

Its national executive committee included John Davis of U.S. Steel, Sewell Avery, president of Montgomery Ward, and Grayson M. P. Murphy, a director of Guaranty Trust. These and six other corporate leaders represented Morgan interests, who were particularly upset with the F.D.R. administration. One year earlier Roosevelt had taken on Morgan and won, humiliating him before a Senate committee investigating Morgan's non-payment of taxes, abandoning the gold standard (leaving Morgan a great loser) and passing a banking act that effectively ended the financial domination of the House of Morgan. More than anyone, Morgan stood to gain from Roosevelt's demise and as much as anyone Morgan was implicated in one of the most bizarre plots in American history, a plot (confirmed by the Dickstein-McCormick House Investigating Committee) to forcibly overthrow the Roosevelt administration and with it, the government of the United States.

The plot fell through when the man selected to lead a million veterans on a march on Washington—General Smedley Butler, a popular hero at odds with Roosevelt—exposed the scheme, which by then had millions of dollars at its disposal along with arms and equipment from Remington Arms (acquired by DuPont shortly before the formation of the League).

On November 20, 1934, General Butler revealed the attempted corporate coup in testimony before a private session of the Special House Committee on Un-American Activities. After completing its investigations, the committee confirmed the attempted fascist coup d'état in an official report to the House on February 15, 1939: "In the last few weeks of the Committee's official life it received evidence showing that certain persons had made an attempt to establish a fascist organization in this country . . . There is no question that these attempts were discussed, were planned, and might have been placed in execution when and if the financial backers deemed expedient."

For reasons unknown, Roosevelt decided not to pursue the affair. Perhaps he feared plunging the country into a deeper depression if business confidence was lost; perhaps he feared the political crisis that would have ensued had he taken legal proceedings against such powerful interests, an action that would have called into question the entire free market system.

Despite the committee's reports, no indictments were handed down.

The head of the American Civil Liberties Union was outraged, noting "the Congressional Committee investigating un-American activities has just reported that the Fascist plot to seize the government . . . was proved, yet not a single participant will be prosecuted under the perfectly plain language of the federal conspiracy act making this a high crime . . . Violence, even to the seizure of the government, is excusable on the part of those whose lofty motives is to preserve the profit system."

Even the press largely ignored this somewhat newsworthy event. Of all the country's large newspapers, most of which were then (and are now) controlled by well-financed syndicates, only four published details of the conspiracy and the corroborating testimonies. The corporate leaders were punished with a stern wag of the finger from the committee's reports: "Armed forces for the purpose of establishing a dictatorship by means of Fascism . . . have no place in this country."

As important as energy and energy-related industries were in global politics before 1945, they have come to be of overriding concern since. Canada's, Germany's, Britain's, and Australia's car markets are all dominated by the Big Three, who also operate or intend to construct major facilities in several politically sensitive areas of the world, including the Soviet Union, East Germany, Egypt, Israel, Saudi Arabia, Korea, Chile, Indonesia, Ireland, and South Africa. At least one U.S. Secretary of Defense has warned that these investments could seriously impair national security.

But it has been oil that has drawn all the world's attention, particularly since 1948, when the U.S. stopped being an exporter of petroleum and became dependent on foreign supplies and multinational suppliers. Within two years, the U.S. would find itself hostage over oil.

After World War II, with financial assistance from the U.S. government, the American oil multinationals were able to secure the concessions of the Saudi oilfields. President Roosevelt noted that if Saudi oil was as vital to U.S. interests as alleged, and if his government was to be putting up the money, his government was entitled to have something to say about the operation. Roosevelt wanted part ownership, and the Petroleum Reserve Corporation was set up to obtain an equity interest. But Roosevelt's plan was blocked by the oil industry and never completed, setting the pattern for future government-industry relations: the government provided diplomatic and financial support in the industry's operations abroad, but had no means of overseeing operations or formulating a policy of its own. The U.S. government became captive to the oil companies, now organized as a consortium called Aramco. Cut off from independent information, America was virtually forced to accept Aramco's word for what was in its national interest.

In 1950 it became in the national interest to subsidize Saudi oil with American tax money. The Saudi rulers had been pressing Aramco for higher royalties from their oil fields. The oil companies managed to persuade the U.S. Treasury to let them pay Saudi Arabia the taxes that would normally have gone to America. The net effect was a U.S. tax holiday for the multinationals, who paid virtually no taxes to their home country. In 1950 Aramco paid $50 million in taxes to the U.S. and $66 million to Saudi Arabia, for a total of $116 million. In 1951, Aramco also paid $116 million in taxes, only now $110 million went to Saudi Arabia and $6 million went to the U.S. The $44 million less that the U.S. received went instead to the Saudis. In future years Aramco would pay nothing to the U.S.

In 1972 Gulf reported that it turned over, in foreign and federal income taxes, $812 million, leaving it with a net income, after taxes, of only $197 million (a tax rate of over 80 percent). Noble as this may seem, that huge sum of taxes did not help fill America's coffers but went to foreign countries—Gulf did not have to pay taxes in the United States of America. The rationale, going back to 1950, had been simply that it was good foreign policy to keep the Saudis happy.

The same rationale—that it was in the U.S. interest to keep

Venezuela happy—resulted in the U.S. oil companies agreeing to higher prices for Venezuelan crude. This time, though, it wasn't the U.S. Treasury but America's friendly neighbors to the north who were made to pay the difference.

Standard's interests went beyond keeping the Venezuelans happy, however. Their concession in Venezuela was about to expire, and they didn't expect the Venezuelans to renew it. Wanting to pump as much Venezuelan crude out of the ground as they could, yet needing to sell it at a higher price than the Americans on the East Coast would tolerate, Standard simply marketed it to Canadians on the East Coast through its Canadian subsidiary, Imperial Oil. Western Canada's oil supplies could then be shipped south, to the U.S. Midwest, instead of to Eastern Canada, which would be supplied by Venezuela. There was no need for Standard to fear that a trans-Canada oil pipeline would be built, the way a trans-Canada natural gas pipeline had been built in 1957. The Canadian oil industry was 77 percent U.S. owned. Canada's Royal Commission on Energy, in rejecting this pipeline, stated candidly that the multinationals would not support such a policy.

Even though the Maritimes were paying more for Venezuelan oil than New England (which received lower priced Middle Eastern oil or "discounted" Venezuelan crude), the arrangement worked to Canada's financial advantage. Alberta's oil was sold to the United States at 25 cents a barrel more than Venezuelan oil cost—Canada was buying cheap and selling dear to the U.S., whose price, until 1973, was $1 to $1.50 higher than world price.

This view of the military—the Chase was seeing it as an aid to trade rather than an instrument of foreign policy—was complemented by a full page ad in the *New York Times* on September 9, 1965, signed by David Rockefeller and the other members of the "Committee for an Effective and Durable Peace in Asia," an organization formed by Chase board members. The U.S. government asked the Chase to open a branch in Saigon to handle the U.S. embassy and military funds.

At stake, of course, was not just a war, but the massive oil reserves off the Chinese and Vietnamese coasts. To win the war, Rockefeller advocated a tax increase of "at least $10 billion" as late as 1968, long after others had quit looking for a military victory. In 1971, still hopeful of victory over the enemy, Rockefeller announced in a major

Singapore speech that the U.S. oil companies were planning on spending $35 million in capital investment along the western rim of the Pacific. After the fall of Saigon in 1975, the Provisional Revolutionary Government of South Vietnam took Rockefeller up on his offer and negotiated with Gulf and other U.S. giants for drilling rights off the Vietnamese coast.

Rockefeller also had kind words to say about the Chinese government, urging a great deal more trade with them while scolding the U.S. for its lack of realism in acting "as if a country of 800 million people did not exist." After his lengthy audience with Chou En-lai in Peking in 1973, the Chase was named correspondent bank for the Bank of China.

This arrangement had nothing to do with the shrewdness of Canadians. Worried about increasing oil imports undermining U.S. foreign policy, Eisenhower imposed voluntary limits in 1957 and then mandatory oil import limits in 1959. But the policy had an overland exemption, because Canadian oil imports were considered "as secure a source as any domestic production." This allowed the multinationals to tap Canada's western oil fields, (saving transportation costs) without Canadian oil affecting the amount of Middle Eastern oil which could be imported into the U.S. (tax free). The multinationals continued to receive higher prices for U.S. domestic production and to pump Venezuelan crude out of the ground while the pumping was good.

Canada had become an important factor for the multinationals since 1947, the year of the oil find at Leduc, Alberta, and the year before U.S. self-sufficiency ended. (Although Imperial Oil had been bought out by Standard before the turn of the century, when it was pumping Southwestern Ontario oil, it wasn't until the Leduc find that Canada made the world oil map.)

Oil supplies were in jeopardy worldwide. Iran nationalized its oilfields in 1951. The Suez Canal was disrupted in 1956. The Biafran war cast doubt over Nigeria's supplies. Revolutions inflamed Iraq, Libya, and Venezuela. These disturbances convinced the multinationals they could not leave politics in the hands of the politicians. They would have to take a more active role in shaping global strategies.

The Iranian Revolution of 1951 was short-lived and, with CIA help, the Shah was reinstalled and the oil company's position secured. But the CIA was only one agency at the multinationals' disposal. Their diplomatic efforts over the last quarter century—many of them conducted through David Rockefeller's Chase Manhattan Bank—are surpassed only by their conception of their place in the world.

The Chase is an oil bank, its business often indistinguishable from foreign policy itself. In its massive dealings in international banking and oil, national policies can not be ignored. It is also lender to the oil companies and trustee for many of the private fortunes (not just the Rockefellers') whose wealth is in oil stocks. Through its prestigious Petroleum Department, the Chase is considered most knowledgeable about loans on proven oil and gas properties in connection with their development, purchase, and merger. At the Chase, Rockefeller chaired a board of directors interlocked with the boards of Exxon, Standard of Indiana, Shell, AT&T, General Foods, Allied Chemicals, Honeywell, and dozens of others. The Chase itself was a leading stockholder in Jersey Standard, Atlantic Richfield, AT&T, IBM, CBS, Motorola, Safeway, and others. Not surprisingly for an oil bank, the Chase strongly emphasizes transportation in its portfolio: it is the leading stockholder in United, Northwest and National Airlines, with large holdings in TWA, Delta, and Braniff, and holds hundreds of millions of dollars in outstanding loans to fourteen major airlines.

Every year Rockefeller hosted a board meeting of the World Bank, many of whose members he had worked with personally at the Chase. Every year, on his dozen or so trips abroad visiting foreign branches or attending monetary conferences, Rockefeller would be accommodated by heads of governments as if he were secretary of state.

The Chase was one of the major creditors to Cuban dictators for half a century, and when the National Security Council decided to invade Cuba, five of those present were David Rockefeller's close friends and associates.

Future co-operative U.S.-Soviet relations were spawned during his 1964 visit to Khruschev. The Chase was later selected by the Soviet government to be the first U.S. bank to open a representative office in Russia. In 1970, Rockefeller hosted Romania's President Nicolae Ceausescu and spoke out in favor of giving Romania "most favored nation" trading status, paving the way for trade with that iron curtain country. The Chase is now the leading correspondent bank for Romania.

Concerned about the business effects of political instability in the Pacific, the Chase strongly supported the decision to bomb North Vietnam and send in troops to the South in 1965. "In the past, foreign investors have been somewhat wary of the overall political prospect for the region," reported the vice president of Chase's Far Eastern operations in 1965. "I must say, though, that the U.S. actions in Vietnam this year—which have demonstrated that the U.S. will continue to give effective protection to the free nations of the region—have considerably reassured both Asian and Western investors. In fact, I see some reason for hope that the same sort of economic growth may take place in the free economies of Asia that took place in Europe after the Truman Doctrine and after NATO provided a protective screen."

In 1970 in the Middle East, while Secretary of State William Rogers was preparing his Mideast accords, Rockefeller told President Nixon of President Nasser's revelation to him—that Egypt would rather deal with the U.S. than the U.S.S.R. To underscore Egypt's intentions after Nasser's death, in March 1971, President Anwar Sadat and his wife were on the front page of Egyptian newspapers, smiling broadly with David Rockefeller and his wife. On Rockefeller's return home from this trip (which included conferences with heads of state in Jordan, Lebanon, and Israel) the Chase proceeded with plans to open a branch in Bahrain.

The Chase opened its Cairo office in 1974 while making an $80 million loan to Egypt for the Suez-Mediterranean pipeline and entering into discussions with King Faisal over what might be done with the huge foreign currency balances Saudi Arabia had accumulated after the OPEC oil increases.

As Iran's and the Shah's personal banker, (whose fortune was estimated at $6 billion), David Rockefeller and the Chase took care of many personal details for the Shah during the turbulent period that led to his fall from power in 1978. In 1979, after mounting an enormous lobbying campaign on his client's behalf, Rockefeller succeeded in bringing the Shah to the United States over President Carter's objections, touching off one of the most incredible political dramas in world history.

The political influence of oil isn't going to die down in the 1980s, and the multinationals wielding that influence won't yield it voluntarily. Nor can we expect multinationals to act in the national interest: they see themselves as governments, and for good reason. Were General Motors, the largest company in the world, to be considered a government, its earnings would rank it about tenth in size among the non-communist countries. The net operating revenues of GM are greater than the gross national products of over 90 percent of the world's countries. Of the next fourteen largest multinationals, seven are petroleum companies (Exxon, Gulf, Texaco, Mobil, Standard of California, Royal Dutch Shell, and BP) with annual sales in excess of the GNPs of countries like Sweden, Germany, India, Poland, and Brazil.

The threat to democracy and democratic institutions is staggering, and of such a magnitude that President Eisenhower used the occasion of his farewell address to the nation to warn Americans with a much misunderstood speech. "In the councils of Government, we must guard against the acquisition of unwarranted influence, whether sought or unsought, by the military-industrial complex. The potential for the disastrous rise of misplaced power exists and will persist. We must never let the weight of this combination endanger our liberties or democratic processes."

Eisenhower, with access to information about the raw role of multinationals during World War II (much of which has only recently been declassified), knew whereof he spoke. As supreme commander of the allied forces, he found himself in the middle of a war waged on both sides by wings of the same multinationals. As supreme commander, he was forced to protect his men and machines from Axis weapons made by General Motors and Ford.

After the war Eisenhower's investigation of I.G. Farben concluded this company was a threat to world peace. He threw his weight behind detailed recommendations that it be dismantled. As victor over the Nazi army, he found himself helpless in destroying one enemy industry. And then as President in the Fifties he saw events unfolding in Iran, in the Suez, in South America—all revolving around an oil industry over which he had little control.

His final act in office—January 17, 1961—was to send out a warning we would be slow to accept. About five years later, Ralph Nader took on General Motors. About ten years later, environmentalists would begin waging an expanded battle, joined by the churches, civil liberty groups, and other organizations across the country. During the 1970s, these individuals and organizations have independently come to a startling realization: in almost every case of dissatisfaction with our democratic system, corporations able to exert monopoly influence are involved.

Those objecting to strip-mining found opposing them a coal monopoly. Those fighting for non-smokers' rights found themselves taking on the tobacco monopoly. The boycott of Nestlés over its sale of baby formula to Africans, the church campaign against banks dealing with South Africa and Chile, the Prairie protests over the patenting of seeds, are all directed against the activities of monopolies. The most publicized campaigns are those mounted against our largest and most powerful monopolies—the interlocked arms, energy and automobile industries. Freedom of choice, and so individual liberties, are curtailed in every case through some form of monopoly manipulation.

Reform in many areas is required. Our efforts can be directed toward correcting the countless existing abuses and the countless abuses to come. Or our effort can be directed toward correcting the monopoly structure, diminishing the ability of corporations to perpetrate any abuse.

# 8

# *Monopolies: Building Them Up, Breaking Them Down*

"MY HOUSE IS buying some land in Florida."

"My daughter's teddy bear has been charged with criminal negligence."

"My umbrella supported the Liberals in the last election. What about yours?"

Outrageous dialogue? Certainly.

Attributing any initiative or reason to material goods is so absurd anyone expressing these thoughts is subject to ridicule. A piece of property cannot take any action or think anything, and with the exception of some schizophrenics, we do not pretend otherwise. In this we are fortunate. To attempt to reconcile the irreconcilable—as a colony of lunatics might—can be easily done. A few plausible assumptions might need to be accepted; some inexplicable phenomena might need to be taken on faith. But as primitive tribes have incorporated scores of superstitions into their customs, beliefs, and even laws, modern lunatics could create a functioning society on false principles without feeling the full consequences until much later.

Luckily our society has spared itself a lot of trouble and avoided treating umbrellas as animate beings. Instead we treat corporations as animate beings. Corporations can buy land, be charged with criminal negligence, support political parties. The evolution of the corporate species has surpassed in speed and scope that of any human or animal species. Unlike nature's evolution, corporate evolution is man-made: it is no more than the progression of our laws to their logical conclusion.

Unless the corporation is given the vote, historians of the future will have great difficulty understanding its status today. In law, corporations are persons with a life of their own, enjoying many of the same rights (and often greater rights) than human persons enjoy.

To prevent individual interests from unduly influencing the political process with large campaign contributions (which can count for many votes), there are strict limits to the amount and manner in which these monies are given and received. This practice was necessary to maintain the principle of "one person, one vote." Yet unlike human beings, who are unable to treat money from their left pockets and their right pockets as independent donations, corporations can funnel money indirectly through affiliates and subsidiaries, and through industry associations set up for the purposes of political lobbying. Corporate persons can live indefinitely; they can be many places at once. Able to own other corporate persons, they indulge in legalized slavery, a power human beings no longer exercise over others. Able to create different categories of stockholders wielding varying amounts of control, corporate persons legally practice discrimination. Able to print their own money by issuing bonds or reissuing stock, corporations hold powers tantamount to those of sovereign states.

To the historian studying Western civilization, these are not frivolous distinctions. Our system of justice and social responsibility is based on property rights—the inviolate rights and responsibilities of ownership. As soon as the distinction between property and the property holder becomes blurred, justice itself becomes blurred.

Rights are no longer related to responsibility. Self-correcting mechanisms break down. Monstrous aberrations can become the norm. But give the corporation the vote—make it a sort of super person in name as well as in fact—and the contradiction disappears. The final stage in the evolution of corporate development will have occurred, assuring a stable corporate future of monopolistically controlled prices. Such a move would make full sense of the corporation's past history. The corporation would then have come full circle, being defined in the end as it was originally conceived—as a monopoly. Corporations were originally nothing more than devices for establishing monopolies, which have a somewhat longer tradition.

Aristotle knew, as did his forefathers, that "the endeavor to secure oneself a monopoly is a general principle of the art of money-making." Though nothing new, monopolies were generally outlawed, as their workings were well understood. "We command that no one may presume to exercise a monopoly of any kind of clothing, or of fish, or of any other thing serving for food, or for any other use," ordered Emperor Zeno in 483 A.D., "nor may any persons combine or agree in unlawful meetings, that different kinds of merchandise may not be sold at a less price than they have agreed among themselves. Workmen and contractors for buildings and all who practice other professions and contractors for baths, are entirely prohibited from agreeing together . . . and if anyone shall presume to practice a monopoly, let his property be forfeited and himself condemned to perpetual exile."

The Magna Carta, in 1215, tried to protect against monopolies by limiting the royal prerogative of granting monopolies. But the practice continued. Elizabeth I rewarded her favorites with patents or monopolies on a scale previously unknown. During her reign, currants, salt, iron, powder, oil of blubber, sulphur, new drapery, dried pilchards, vinegar, cloth, cards, saltpeter, sea-coals, lead, oil, glass, paper, starch, tin, *aqua vitae,* brushes, pots, bottles, calamine-stone, calfskin, fells, pouldavies, ox-shin bones, soap, train oil, potashes, anise seeds, the rights to transport iron ordnance, beer, horn, and leather, and the rights of importing into England Spanish wool and Irish yarn were all monopolized, leading one parliamentarian to shout out in the House, "Is not bread in the number?"

"Bread!" the other parliamentarians cried out in astonishment.

"Yes, I arrive you," he replied. "If affairs go on at this rate we shall have bread reduced to a monopoly before the next parliament."

Because of the outcry the monopolies were canceled, it being noted that monopoly grants had three inseparable attributes: the price of the thing monopolized was raised, its quality deteriorated, and the grant tended to the impoverishment of competitors.

Patents—originally given as royal favors unrelated to any talent possessed by the recipients—were still granted, but only to reward and encourage invention, and only for a term of fourteen years. Patents became gifts from the state, in recognition of service to the state: contractual arrangements as opposed to inherent rights. Unlike works of art, which are literally unique products, inventions more closely resemble ideas that are in the air—not creations but discoveries often made simultaneously and independently by different people in different parts of the world. There was only one source of Picasso's paintings or Shakespeare's plays, but Germans and Canadians both claim the telephone, the French and Americans both claim the motion picture camera, and almost everyone claims the automobile.

While patent monopolies were under attack a new form of monopoly was established—the corporation in the form of great trading companies like the East India Company, the Virginia Company, and the Hudson's Bay Company. Though blamed for the ultimate decline of Britain's foreign trade because they barred the rest of the country from trading with three-quarters of the known world, by today's standards these early corporations (of which only Hudson's Bay survives) had limited mandates. Corporate monopolies were considered necessary evils by governments unable to convince businessmen to take great risks without greater rewards, and the corporate charter strictly limited rewards to the area in which the risk was taken.

The British philosophy carried through to Britain's North American colonies where, well into the nineteenth century, corporations were rarely allowed to aggregate more than $1 million, or to stray past their territorial boundaries, and so were denied the right to do business or own property outside their own area. Corporations could only owe so much, or exist for so long, generally twenty, thirty or fifty years. Far from today's *carte blanche,* charters specified a single purpose, or a limited number of purposes, such as a railroad operation or a mining and manufacturing project. The idea of one corporation holding stock in another was unknown.

Shareholders had the power of directing corporate policy; any proposal to change the corporation's assets, capitalization, share structure or bylaws needed unanimous approval. The shareholders meeting—akin in atmosphere to lively town meetings—was the critical decision-making forum, choosing directors by majority vote and able to replace them at will.

Though monopolies still existed, these limits on the corporation effectively prevented a proliferation of monopolies. Even the monopolies that emerged, such as Standard Oil's, were insecure operations, depending on some good will from the politicians in their pay and much good will from co-conspirators. As long as Standard's operations were limited to New York, Pennsylvania, and Ohio, Rockefeller felt he could control the few other refiners in his cartel. The further afield he ventured, the more he became dependent on the co-operation of potential competitors.

Rockefeller devised the Trust in order to cross state lines and maintain full control. But even this arrangement was kept under strick secrecy and needed to be smoothed over with bribes of entire legislatures. When the Standard Trust, and other trusts that had sprung up in everything from whiskey to lead, were taken to court in the 1890s and ordered broken up, the dissolution was not due to the new anti-trust statutes but to the common law of corporations. This law simply ruled that a trust was "ultra vires [beyond the powers] of the corporation and against the public policy." Of all the prosecuted national trusts, the Standard was the only one to survive, doing so in blatant defiance of the law. For five years Standard simply ignored the Supreme Court of Ohio's order to break up, and when the state attorney general took action to revoke its charter, Standard left for New Jersey, a paradise of corporate prostitution.

As documents in the state archives indicate, no lesser figures than the governor, the secretary of state, and other leading politicians in New Jersey were bribed. The money was well spent, however; New Jersey held nothing back from its new clients. Just one year after the Sherman Anti-Trust Act of 1890, New Jersey introduced legislation that effectively reincarnated the trust organization by authorizing a single corporation—called a holding company—to control the stock or assets of its competitors in the same fashion as a trust. In 1892 New Jersey repealed its anti-trust law.

New Jersey, in effect, had gone into the chartering business, acting as a haven for the misfits and unfit of the business world. In much the same way Liberia now registers the seaworthiness of any ship, New Jersey relaxed all standards, authorizing corporations to buy and sell the stock or property of other corporations and to issue their own stock in payment.

Protection of investors against overpriced stock was ended. The state declared that "the judgment of directors as to the value of property purchased shall be conclusive." Corporations could now buy out competitors without needing to raise a penny in cash, offering them sizable amounts of stock that, without their knowledge, had been even more sizably watered down. Public investors, meanwhile, were being conned into paying cash for shares in prestigious corporations whose assets were often imaginary. New Jersey stock promoters weren't required to verify the worth of the paper they were peddling.

Fifty-year limitations on corporations were removed. Any corporation could now "purchase, hold, sell, assign, transfer, mortgage, pledge or otherwise dispose of the shares . . . or any bonds, securities, or evidences of indebtedness created by any other corporation or corporations of this or any other state, and while owners of such stock may exercise all the rights, powers, and privileges of ownership, including the right to vote thereon." No limit to corporate size, no limit on market concentration. Corporations could carry on business anywhere in the world, merge or consolidate at will. Consolidated corporations could then sell, mortgage, lease, or franchise any property so obtained.

Stockholders' control (and with it stockholder accountability) became a forgotten concept. Different classes of stocks and stockholders emerged: voting and non-voting, common and preferred. Corporations could now sell non-voting shares as fund-raising schemes while retaining the voting shares and powers for themselves, divorcing ownership from responsibility and reducing stocks to the level of lotteries, or play money from some boardroom game.

Directors no longer needed stockholder approval for bylaw changes—New Jersey permitted them to amend bylaws on their own. Nor did they need worry about losing control over their companies: with the use of the proxy vote, their self-perpetuation was assured.

The holding company—believed to have been devised by a J. P. Morgan lawyer in order to create a transit utility in 1893—was followed by Morgan's imaginative use of the "voting trust," as the U.S. government's Money Trust investigation would reveal in 1913: "Thus, on the reorganization of the Southern Railway Co.—by J. P. Morgan & Company in 1894, a majority of its stock was placed in a voting trust which deprived the stockholder of all representation and voting powers and vested the absolute control of the company in the trustees, J. P. Morgan, George F. Baker, and Charles Lanier, who, upon the transfer of the stock into their names, issued the usual trust certificates, which were listed and traded on the exchange instead of the stock certificates."

These new certificates were neither bonds, backed by collateral, nor equities representing some form of distant control. They simply represented a vote of confidence by the holder that he was in good hands in the Morgan trusteeship.

For shareholders who might not willingly extend that vote, a life was set to the old certificates, forcing the holder to turn them in for new versions of this privately printed paper currency. Holders of 183,938 original shares rebelled, only to discover, revealed the Money Trust investigation, they were holding virtually valueless certificates: "The result was that those not assenting to the extension of the trust and hence not taking the new trust certificates found themselves with a security not listed on the exchange and therefore without a ready market and not available as collateral."

By 1896 the corporate coup was complete. New Jersey had revolutionized the corporation in order to attract business and the strategy was stunningly successful. Between 1880 and 1896 only fifteen corporations with $20 million in authorized capital had been chartered there; between 1897 and 1904 New Jersey attracted 104. In 1896 alone, New Jersey granted a total of 854 charters for $857,000 in franchise tax revenues. By 1900, 95 percent of all the major corporations were chartered in New Jersey, including Standard Oil, which was about to lose its Ohio charter.

In just a handful of years the concept of the corporation had been turned inside out, a profound change that soon debased the corporate policy of every government in North America. For though some tried to maintain their principles for chartering, the flight of revenue to New Jersey forced all, bit by bit, to "compete" with New Jersey by adopting progressively laxer policies. New Jersey for years would be accused of "trafficking in treason," but its standards became the North American norm. Government efforts to regulate business's burgeoning powers to this day maintain a remarkably dismal record. This despite complacency-shaking revelations from incontrovertible testimony: "Do you think that [further concentration] would be dangerous?" asked Samuel Untermeyer for the government in the famous Money Trust Investigation of 1912. Testifying was George Baker who, along with his associates, J. P. Morgan and James Stillman, were the dominant forces in the centralization of finance and industry, holding 341 directorships in 112 dominant corporations with aggregate resources or capitalization of $22.245 billion.

> BAKER: I think it has gone far enough.
>
> UNTERMEYER: You think it would be dangerous to go further?
>
> BAKER: It might not be dangerous, but still it has gone about far enough. In good hands, I do not see that it would do any harm. If it got into bad hands, it would be very bad.
>
> UNTERMEYER: If it got into bad hands it would wreck the country?
>
> BAKER: Yes, but I do not believe it could get into bad hands.
>
> UNTERMEYER: So that the safety, if you think there is safety in the situation, really lies in the personnel [sic] of the men?
>
> BAKER: Very much.
>
> UNTERMEYER: Do you think that is a comfortable situation for a great country to be in?
>
> BAKER: Not entirely.

The sovereignty of governments is now suspect; their ability to be a check and balance to corporate autarchy is in serious question. When the Canadian government, partly due to post-OPEC public pressure over the role of oil giants in the economy, decided to set up a Royal Commission on Corporate Concentration to investigate monopolies, it named the president of a multinational oil company and a corporate tax lawyer as commissioners. The final report they wrote in 1978 literally spoke volumes about the place corporations have in our society. Not only does the Canadian economy need to be further monopolized and corporate income tax and capital gains tax abolished, the commissioners decided, but the role of the corporation needs to be redefined. On dealing with the government: "A government 'policy' is often no more than the expression of a hope or sentiment, and a particular policy or an aspect of it frequently conflicts with another. It will seldom be possible to draw useful conclusions about corporate social responsiblity by weighing a corporation's actions against government policy."

Their conclusion here, that there's no such thing as government policy—certainly nothing worth complying with when corporate interests dictate otherwise—is a theme that's spelled out in their definition of the public interest: "A corporation, or indeed any institution, has no single 'public interest,' but rather has many publics, with many competing interests. The 'publics' of a corporation include shareholders, creditors, employees, customers, suppliers, governments and local, national, and even international communities . . . We do not imply in this *Report* that there is any best allocation of corporate resources among competing publics . . . We make no a priori assumption that any given allocation is superior from a public interest sense."

The corporate interest, in other words, may be greater than the national (or even international) interest. The corporate imperative—stated here in theory—has often been followed through in practice: during the Second World War, with the questionable behavior of Standard Oil, I.G. Farben, GM, and Ford; since the war with the machinations of the oil giants. Corporations serve their multinational interest with great equanimity, as witnessed by INCO and its Guatemalan connections, by the direct role our banks are playing in propping up the South African regime, by ITT's role in toppling the Allende government in Chile.

Because of the potential conflicts of interest, national laws have always tried to restrain the power of business interests. Written right into the U.S. Constitution is protection against monopolistic practices, with patents promoting "the progress of science" secured only "for limited times." Canada, giving out patents before Confederation, stressed that society wanted something back immediately by requiring the "new invention . . . so far as is possible be worked on a commercial scale in Canada without due delay."

In fighting the trusts, the interrelated economies of Canada and the U.S. responded almost in tandem. The first federal Canadian anti-trust act was passed in 1889, just one year before the U.S. Sherman Anti-Trust Act, and was just as toothless. Major additional anti-trust legislation came to Canada again in 1910 and 1935, largely paralleled by U.S. legislation in 1914 and 1936. Minor legislation, relentless regulation, and endless investigation have marked the periods before, between, and since, always to no ultimate avail, even when apparently successful, and always at great expense and great confusion. Society has been put through the corporate hoop.

First we tried to outlaw restraint of trade through price-fixing, boycotts, or dividing markets. Then we made it illegal to monopolize or conspire to monopolize. Then, to nip monopolies in the bud, we tried to restrict practices involving different prices for different customers, exclusive dealerships and dealings, reciprocal arrangements, mergers and interlocking directorates. Unable to prove the existence of monopolies, we then tried to define monopolies as existing when the four largest companies in an industry control 50 percent or more of the market, or when any one corporation maintains an average rate of return in excess of 15 percent over a period of five consecutive years out of the seven most recent. Then it looked as if we should control advertising that helps perpetuate monopolies. A final attempt, introduced in the Canadian Parliament in 1977 and then stalled, would outlaw monopolies, even inadvertent monopolies that arose without any conspiracy by any party.

Ironically, even should every loophole in every anti-trust law be closed, the public could not be assured that the laws would be enforced. Corporations can pit their size against the size of the judicial system, and frustrate government attempts at regulation. When the U.S. government contemplated taking IBM to court in the late 1970s the company estimated the preparation for its defense would require gathering 5 billion documents, would take 62,000 man-years of effort, and would cost $1 billion. (For IBM to get a fair trial, in other words, would involve the mobilization of 6,200 employees for ten years—or 62,000 employees for one year—at a cost of $1 billion.)

For the judge to weigh the evidence fairly would be another matter. To read 5 billion documents, even at an average of one minute per document working twenty-four hours a day, 365 days a year, the judge would need to live 10,000 years.

There can be no justice for IBM if the judge does not consider all relevant evidence, and no justice for society if the judge does. Little wonder, then, that major anti-trust suits consume an average of eight or nine years. Too often a case is abandoned because the resources of government fall short.

The modern business corporation has emerged as the first institution to claim significant and unchecked powers since the decline of the church and the supremacy of the nation state in the sixteenth and seventeenth centuries. As powerful private governments, with the economic clout of nation states and populations to match, multinational corporations have acquired power that is incompatible with the notion of a free-enterprise democracy. Either corporate power or Western democracy will have to give way. If democracy is to acquire supremacy, the principles of a democracy, which include individual rights based on property rights, must be respected. For this to happen, the bizarre notion of the corporate citizen must be retired. The corporation must be rationalized as property by stripping it of its human rights and responsibilities and returning them to its human shareholders.

Once the corporation loses its personality it stops being a fearsome institution of oppression and starts to become a useful organizational tool appropriate in marshaling diverse interests together for the purpose of commerce in a competitive free market.

Nothing revolutionized the corporation more than the decision to treat it also as a person. Though it acquired these "extra-corporal" powers gradually, an 1886 U.S. Supreme Court decision confirmed the corporation was a "citizen." The Canadian law against combines was profoundly influenced by the American evolution of the corporation. Prime Minister Mackenzie King (who, ironically, had been a Rockefeller employee during an American sabattical from politics) emphasized that the anti-monopoly measure "does not propose to place the parties in a position of defendants in a criminal court but treats them as persons whose business for the time being is being examined."

The revolution was heightened and confused by an incredible oversight: the corporation retained its status as property, making it both person and property, the owner and the owned. Paradoxically, it was able (as property) to demand its rights of itself (as owner)—answerable, in effect, to itself.

Financier J. P. Morgan, whose holding company controlled anthracite coal mining in America, ably demonstrated this flexibility as far back as the turn of the century. During a coal strike that was crippling the country Morgan was asked to accept arbitration and end the strike but he professed powerlessness, explaining he could not speak for the presidents of the individual coal mining companies. Morgan was doing no more, here, than upholding the rights of the companies as corporate citizens. The presidents of the companies, on the other hand, could not act without the consent of their shareholders—the owners on whose behalf they were acting. But these same shareholders were the Morgan interests. Despite intense public pressure, and personal intervention by President Teddy

Roosevelt, the coal magnates (according to Roosevelt) "insolently" declined, refusing "to consider the public had any rights in the matter." Only after Roosevelt ordered military plans drawn up for government seizure of the mines did Morgan recognize his responsibilities and go along with compulsory arbitration.

Morgan, who scandalized the public by saying "I owe the public nothing," was only expressing conventional corporate philosophy. Vanderbilt's parallel sentiment that "the public be damned" was not the mere expression of a preference but contained a more profound inference of the principles by which corporations were governed. Vanderbilt also expressed a more practical principle: "What do I care for law? Hain't I got the power?"

Determining responsibility for an individual's actions is difficult enough—there are often so many ambiguities and extenuating circumstances that even a judge and jury rarely feel certain the right decision has been made. Under the corporate cloak, the problem of assigning responsibility for questionable behavior is compounded, and the penalties available are often meaningless. Corporations can yawn at proscriptions traditionally designed to strike terror in those made of flesh and bones. Zeno punished monopolists with the social penalty of banishment; what would it mean to banish a corporation? An individual performing GM's functions for the Nazis during World War II would have been shot; a corporate charter doesn't bleed. To a human, the social opprobrium that results from committing even a petty crime, such as shoplifting, often far outweighs the fine that may be levied. Corporations don't have trouble sleeping at night when found guilty of far worse crimes than petty theft; their employees and shareholders may not even know crimes were committed. An executive leading his corporation astray need not worry about ever having to say he's sorry with his pocketbook. To overcome the threat of fines, corporations routinely take out liability insurance for their officers and directors, with the premiums paid for by the corporation. Shareholders are not even financially responsible for their corporation's failings. Even should strong corporate penalties be devised, the wrong people would be punished. When an individual's actions lead to a court fine a year later, he must pay the penalty for his own misdeeds. When a corporation's actions lead to a later fine, the shareholders who must pay are not necessarily those presumably responsible (as owners) for the crime. (In fact, in cases where major shareholders with inside information know a large penalty may be imposed they can conveniently—if illegally—dispose of their shares and their responsibility.)

Corporate contradictions have managed to completely obscure the relationships between the rights and responsibilities of ownership upon which our system of justice is based. The obscurity has become more pronounced with the evolution of the corporate freedom to create wholly owned subsidiaries needing no human shareholders to distort corporate thinking. Such companies can operate with nothing but a single corporate owner and a single officer who can be kept in legal ignorance, able to take orders from above without knowing why, yet not fully responsible for his actions unless he knows why. No assets, no employees, nothing else but an address is required. Wholly owned subsidiaries have been able to be parent to wholly owned corporations of their own and these, in turn, have been able to gain control of the grandparent organization. Corporate law has allowed corporations to control themselves, and statistically it is likely that the ultimate absurdity has occurred—that corporations have already broken off from human shareholder control, influencing public policy as corporate citizens without the direct or indirect involvement of any human owners.

Confirming the corporation's bizarre independence is a 1980 U.S. Supreme Court decision allowing corporations to patent new forms of life. Should genetic engineering succeed in cloning human forms, corporations would be in a position to create human beings to their requirements, perhaps improving on them. Cloned humans could then become shareholders in their corporate creators, or vote on election day if corporations had not yet won this privilege.

Antisocial behavior aside, one of the desired functions of a corporation is the shunting of responsibility. To encourage speculation, corporations have been given limited liability—there's only so much their shareholders can lose. A corporation with $1 million in debts and a limited liability of $50,000 can declare bankruptcy and force its suppliers and creditors to pick up the tab for the remaining $950,000. Responsibility has been shunted from the owners of the property to those who trust it. In the case of the individual who does not want to risk his house in entering a business venture, the limited liability that incorporation offers can be seen as a form of social insurance. But corporations have parlayed this limited liability for individuals into unlimited liability for society. Corporations, vehicles designed to take chances, have decided to protect themselves by setting up subsidiaries to take the risks for them. Supertanker oil spills can cause such heavy damages, for example, that oil companies typically incorporate each vessel as a separate company. When a *Torrey Canyon* or an *Amoco Cadiz* cause millions of dollars in damage, the parent companies are not liable for the damages—the corporate vessel can be declared bankrupt, the damage not covered by insurance being borne by victims of the spill or by the environment.

On a smaller scale this shunting of responsibility has become a routine occurrence. Even small corporations deciding to expand into risky ventures prefer to do so with the investment dollars of others. The house and home of the individual investor is no longer being protected through the device of a corporate charter—it is now the corporate citizen who is being protected by shunting risks onto small suppliers and individual investors. Corporate subsidiaries encourage a speculative society in which those with the most information about a venture's prospects are those least willing to assume its risks. Individual responsibility, and so individual rights, become severely degraded.

Remove the right of citizenship and the corporation can become compatible with a democratically run society. As a piece of property owned by others, corporations, like houses or hobby horses, could not own other properties. Holding companies, conglomerates, wholly owned corporations, subsidiaries, branch plants, affiliates would disappear, all becoming companies directly owned by human persons. Shareholders in a former conglomerate could still have interests in its many constituent companies, but the interests would be through direct ownership, carrying the direct responsibilities of an owner, without obscured lines of responsibility. In 1968 Standard Oil of New Jersey (now Exxon) had investments of about $1 billion, yet it was not possible, found one scholar, "to identify from public sources even the names of all the corporations in which it had investments, and it was not possible to identify the value of the assets of any of these corporations."

Though under this system corporations would not be able to own other corporations, legitimate expansion could take place in the same way people buy assets from other people. A corporation could buy out the assets of other corporations—their land, their buildings, their clients, their furniture . . . everything but their charter. This limitation would be more than symbolic. When corporations are bought out, former shareholders typically get paid in shares of the new corporation that's created, allowing companies to use, in effect, the assets of the bought corporation to do the buying. Shareholders receive their own money in payment. Also, because there has been no actual sale of goods but rather a union of corporate persons, sales taxes are avoided at the time of acquisition, and other taxes can be avoided in future.

Remove the right of dual citizenship, and foreign-owned businesses (as well as corporations) become repatriated. A citizen of one country cannot live and work indefinitely in another without losing some rights of citizenship. Yet when it is an American company that lives and works in Canada, instead of an American person, the company gets preferential treatment on both sides of the border. Branch plant operations have become the multinationals' preferred means of carrying on international trade. Instead, foreign trade can take place directly, allowing importing companies the freedom to shop where they see fit.

The tax-saving preference of selling imported goods under a domestic guise can be dispensed with, and the importing-exporting business (under attack since Rockefeller eliminated his independently owned foreign affiliates in the last century) can regain its status as a worldwide entrepreneurial industry. Goods would be available when entrepreneurs wanted to import them, not when multinationals decided to market them.

If the right of citizenship were removed, corporations, as pieces of property, could not make political donations. Businesses, as institutions promoting their particular philosophy, would still be entitled to their many political views. But, as with the modern separation of church and state, the business ideology could not be easily promoted with the company's own tax-deductible dollars, but through tax-deductible donations from its individual owners or shareholders.

Pieces of property could not be recipients of the state's largesse in return for meritorious services performed. A medal for bravery can't be pinned to a corporation's chest. Neither can the bravery of one soldier be attributed to someone else. Likewise the patent should be a non-transferable privilege reserved for human persons. Patent holders can market their inventions themselves, and benefit from those who market their inventions for them (as authors can retain the copyright on the books their publishers distribute). But until the patent expires, extraordinary monopoly benefits should be conferred on none but those who have enhanced the state.

Property could not act as a self-investor, an unaccountable reservoir in which to sink huge excesses of cash, either as a dodge to avoid paying personal income taxes, or as an instrument of empire building. Corporate profits should be returned to shareholders in the form of dividends each year so that shareholders can show those earnings along with any others to the tax man. A corporation could reduce the size of those dividends by expanding its operations or it could expand its operations by adding to its normal borrowing requirement at whatever interest rate the rest of society gets. A corporation would no longer be able to insulate itself entirely from the marketplace (as humans can) with a huge reserve. When sales were off, it would have to cut prices; when a competitor came up with a better product, it would be unable to use the monopoly profits of earlier years to price him out of business before jacking up prices again.

While corporations would be worse off when they had to borrow at high interest rates, their shareholders would be better off—they would have the cash on hand to either reduce their personal borrowing requirements, invest in a more productive concern, or reinvest with the corporation when they're pleased with its performance.

Shareholders of large corporations would especially benefit, since large corporations not vulnerable to takeover are typically undervalued—their shares often selling for less than half their worth. Texaco, for example, sells for 28 percent of its asset value and Exxon for 40 percent (while a small oil company like Crystal sells for 150 percent of the value of its holdings). Unless Texaco or Exxon have a buyer, their value is a theoretical one that may never be claimed. (This explains why, after a dissolution, the value of the new companies is greater than the undivided corporation, and why a company's shares invariably rise when a takeover bid is made.)

And, of course, all of society would be better off. Whether the shareholders invested in a business or in a bank savings account, the huge infusions of freed money into the market would ease capital shortages and lower interest rates.

Wealth as a result of good corporate decisions would now go to the owners of that wealth, sometimes resulting in filthy rich individuals instead of filthy rich corporations. But the individuals would be taxed on their earnings; private fortunes would be divided among the heirs and the government (through succession taxes). Unlike corporate fortunes, which can remain intact forever, private fortunes would last about as long as the persons holding those fortunes. The rich do not get richer under our present system, only more powerful through their ability to manipulate corporate fortunes. The combined branches of the Rockefeller family today are worth about $500 million—a large amount next to the minimum wage earner's bankbook but a small fraction of the estimated $1 billion personal fortune (in 1910 dollars) John D. Rockefeller alone had accumulated. The influence of a David Rockefeller today comes not from his share of the $500 million but from corporate rules that allow the power that comes with immense concentrations of wealth to be in the hands of non-owners.

Without these corporate rules, $500 million would not go far—all the Rockefeller family could own outright would be either one of the larger distillers or one of the smaller oil companies.

The corporation, as property, should not be able to exercise legal discrimination by allowing some shareholders to vote and others only to invest money with no obligations as property holders. Those who wish to be free of responsibility can purchase corporate bonds (which will show up as a liability in the corporate books). All property holders should get the same one vote for each share they hold (eliminating situations such as Laurence Rockefeller's control of the $1 billion Eastern Airlines through his holding of 100 percent of the preferred shares—valued at less than $1 million). All property holders should have the same complete access to the records of the corporations they own as does the management that serves at the owner's pleasure.

Once owners can look into the corporate books for themselves they won't need to rely on their own management's opinion as to whether it has been doing a good job, or entering promising new markets, or whether it should be rehired—all the gaffes, poor investments, and dubious policies will be there to appraise along with the good ones.

The annual reports of corporations—whether the corporations are about to go under or not—are currently public-relations pamphlets that tend to paint the same glowing portrait of prosperity. Though financial statements must be certified by outside auditors, the accounting profession has been stripped of much independence and now amounts to little more than a rubber stamp applied at a client's call. According to a partner in one of the continent's largest accounting firms, "my profession appears to regard a set of financial statements as a roulette wheel to the public investor—and it is his tough luck if he doesn't understand the risks that we inject into the accounting reports." This explains the falsification of records that occurred when Atlantic Acceptance, one of Canada's financial giants, went bankrupt to everyone's surprise, how Peat, Marwick could have failed to report the troubles of the Penn Central before it went into receivership, why Price, Waterhouse lent its stamp of approval to large corporations that would soon embarrass it by going out of business.

Along with the property holders' right to information should go the ability to use that information for the good of the company—especially at annual meetings. At present, management—using shareholder money—decides what information shareholders receive. For an individual shareholder to inform co-owners of additional information is often virtually impossible without management's approval—to send a single letter to the 75,000 shareholders in a multinational like INCO could cost $20,000 or more in paper and postage alone. In any instance where a claim by management is unsubstantiated a shareholder should be able—at his company's expense—to inform other owners of his concerns about management's capacity to continue in its function. The effect of this requirement will be to create managements whose public ethics are beyond reproach and whose financial statements are as clean as those of most charities. The corporate ethic, of course, will stay the same, but from the management's point of view, forthrightness will be cheaper than money spent in perpetual internal squabbles. An ever-vigilant spirit will pervade managements who will know they'll be re-elected not on past laurels but on present performances.

That corporate secrets will escape from shareholders to competitors need not be a concern—a genuine commitment to a free enterprise system necessarily precludes corporate secrecy. When the Federal Trade Commission's chief economist said that "contrary to the assertions of many corporations, corporate secrecy—not corporate disclosure—is the great enemy of a market economy in a free society," he was only stating a basic tenet espoused since Adam Smith. The father of economics recognized that the health of the competitive market depends on the unfettered knowledge purchasers have of sellers and competitors have of each other. Once business stops competing on the basis of cost, service or product quality, the law of supply and demand is supplanted by "charge whatever the market will bear."

Corporate secrecy does not help to secure legitimate property rights, but only to make insecure (or to steal) the property rights of others. Every invention, every process, is patentable—a privilege the state is pleased to confer for a limited number of years on all who enrich society with their innovations. The other half of the bargain is that after the patent runs out it becomes public property, available to all. Secrecy is no more than a guise to hide another's legitimate property rights, or to hide some monopolistic practices such as unlisted rebates or monopolistic overcharges. In a free society, the only competitive edge companies could maintain would be in their production or marketing efficiencies. The challenge to others should be in finding a way of doing the job better, not in better hiding corporate secrets.

Corporate secrecy can be safely abolished as soon as patent laws are able to provide protection for property. To date, patents have been turned into a major instrument of abuse.

After completing its final investigation, The U.S. Temporary National Economic Committee stated: "No one can read the testimony developed before this committee on patents without coming to the realization that in many important segments of our economy the privilege accorded by the patent monopoly has been shamefully abused. It is there revealed in striking fashion that the privilege given has not been used, as was intended by the framers of the Constitution and by Congress 'to promote the progress of science and the useful arts' but rather for purposes completely at variance with that high ideal. It has been used as a device to control whole industries, to suppress competition, to restrict output, to enhance prices, to suppress inventions and to discourage inventiveness."

Our industrial history is replete with examples of industries dominated by a few monopoly groups whose power rested on patents. In electrical appliances and electrical equipment, in the communications and electronics fields, in explosives, in synthetic rubber, in dyestuffs, in magnesium, in medicines, in spectacles and optical equipment, in glass containers, in vitamins, in building materials and in explosives (among other prominent examples) the development of the industry was guided not by free market principles but by the arbitrary discretion of groups controlling concentrated patent structures—deciding who gets permission to manufacture, buy and sell, at what price and in which market . . . in effect regulating industries by taking on the role of private governments.

Some remarkably candid corporate memos from the glass industry have come to light. In them the advantages available to corporate patent holders are made clear: "In taking out patents we have three main purposes: (a) To cover the actual machines which we are putting out and prevent duplication of them . . . (b) to block the development of machinery which might be constructed by others for the same purpose as our machines, using alternative means; (c) to secure patents as possible improvements of competing machines so as to 'fence in' those and prevent their reaching an improved stage

Once in control of the patents, a model licensing policy for monopolists is implemented: "Consequently we adopted the policy which we have followed ever since, of restricted licensing. That is to say (a) we licensed the machines only to selected manufacturers of the better type, refusing many licenses whom we thought would be price cutters and (b) we restricted their fields of manufacturing, in each case, to certain specific articles with the idea of preventing too much competition. (c) In order to retain more complete control of the situation, we retained title to the machines and simply leased them for a definite period of years, usually 8 or 10 years, with the privilege of renewal . . ."

The system permitted practically no competition, and, as in the electrical industry, outsiders attempting to enter the industry without first surrendering their patents faced a few years in the courts.

The threat of expensive and protracted patent suits has always been the most effective means employed by monopolies to enforce their private rule over upstarts who would compete with them. To an independent inventor, possession of a patent is little more than an invitation to predatory litigation. Even Edison complained about patent litigation, stating that "when [an invention] is introduced our beautiful laws and court procedure are used by predatory commercialism to ruin the inventor. They don't leave him even enough to start a new invention."

The problem lies not in the principle of protecting property rights through patents but in the restrictions placed on patent holders. Suppose an inventor devises an incredible machine never before conceived—say a machine that converts kitchen garbage scraps directly into usable household electricity. Although the machine would be patentable (so that no one else could use it) in all likelihood the inventor would be unable to market it, because there would have been a score of other patents on various components that went into his design. His alternatives would be to either negotiate with the monopolies holding those patents for permission to share in the "patent pool" that controls new inventions, or to wait out a dozen years or more until the various patents expired and then face years more in court trying to prove that his legitimate achievement does not infringe on another firm's rightful property. The entrepreneurial inventor isn't the only loser. Society as a whole is deprived of a useful invention which might well not have otherwise been invented for years.

When governments grant patents that restrict the patent holders from immediately putting their products on the market, they are infringing upon the inventor's property rights, restricting the free market, and penalizing the public. All new inventions, whether they incorporate old ideas or not, should be free to enter the competitive marketplace. The property rights of the "patent pool" can be safeguarded through the same royalty or licensing system the pool has already adopted for its participating members.

But rather than allowing the patent holders already in the pool to decide who can compete in the marketplace, any enterprise should be allowed to use any patent upon paying the same fee charged others. The music industry has long functioned this way. Any singer is free to record any song without previously negotiating with the songwriter. Where there is dispute over the amount that should be paid, let the suit come after production instead of before, so that royalties can be paid out of the new concern's profits, and the public can benefit from the product and the competition, without any patent holder losing the remuneration that is his due.

In the period after Edison invented the incandescent lightbulb, dozens of competitors, improving on his original design, forced him to further improvements resulting in generation after generation of superior bulbs at ever lower prices. The suits for infringement of patents, for the most part, came afterward, when the electric lamp industry organized itself into a monopoly during the period of the trusts, extracting more excessive profits over a longer period of time than for any other household, commercial, or industrial necessity.

The present morass of business legislation and regulation can be eliminated with this non-restrictive approach to patent-ownership, accompanied by adherence to elementary principles that should be basic to free societies—no secrecy in business and a strict upholding of property rights. By bringing corporations back under control of shareholders, man-made monopolies can be ended, responsibility restored, industry can be deregulated and enterprise freed. Under this system none of the great frauds of the past could have occurred. The secret rebates in the South Improvement Company scheme could not have been kept secret. Morgan's great voting trusts could not have been continued. The institution of blackmail and graft could have been eliminated. Multinational corporations would no longer exist; one corporation could no longer own another. Bureaucracies, like the Foreign Investment Review Agency, designed to regulate foreign takeovers of Canadian corporations, would disappear for lack of work.

Well entrenched existing monopolies could simply be legislated out of existence, or free enterprise principles applied to let them author their own demise. Their demonopolization need not be disruptive; neither need it take long.

For a massive corporation like General Motors, the first step would be to sell off its various subsidiaries and corporate holdings (but without buying new ones) as it normally does anyway in the course of business. Generally, the sale or breakup of conglomerates leads to larger rewards for shareholders. The proceeds from these sales would be passed on to shareholders rather than retained as earnings.

All shareholders who owned non-voting shares would be given the option of selling them or transferring them to voting shares in either GM or its former subsidiaries. All corporations would now be unable to buy out other corporations with their own stock. Economic imperialism, like Great Power imperialism, would belong to a bygone era.

With the abolition of corporate secrecy and the rise of shareholder control, all the magic of monopolists' manipulations will evaporate: reciprocity, exclusivity, price-fixing, rebates, et al will be exposed as soon as suspected. The weak divisions of General Motors will no longer be kept invisible by the support of stronger divisions. Competitors, seeing where the profits are to be made, will be able to target for those markets, forcing GM to give up its monopoly profits in its strong divisions. Unable to support the weak divisions with these funds, GM will have to sell them to others who may be able to do better. Meanwhile, among the million-plus remaining newly enlightened GM shareholders, different ownership philosophies will emerge. Factions will form that will lead to the election of representative boards of directors having members with divergent views. Some divisions will split off as separate enterprises; others will forge tighter links.

Poor business decisions will be publicized to the shareholders; good management will replace bad. The ultimate size of the corporation will ultimately be determined by the market. GM will devolve to the only efficient scale that a freed automotive industry would support—a scale determined not by subsidy or regulation but by the rules of the unsubsidized, deregulated marketplace.

At this scale the nature of decision-making will be based more on the human mentality of the people making the decisions, less on the corporate mentality of the corporate structure that shapes all aspects of present decision-making. That John Z. DeLorean, formerly the GM executive in charge of all car and truck operations in North America, frankly admits that the Corvair was approved "even though serious questions were raised about its engineering" may be sensational but it is hardly significant. That he admits GM executives had little corporate choice in the matter is highly significant: "these were not immoral men who were bringing out this car . . . But those same men, in a business atmosphere where everything is reduced to terms of costs, profits, goals and production deadlines, were able as a group to approve a product most of them would not have considered approving as individuals."

This corporate schizophrenia has done much more than produce unsafe vehicles and an unsound economy. This schizophrenia is one of the least recognized but most insidious agents of social change.

# PART 2

# Scorched Earth Policies

# 9
# *Trafficking*

THE FUTURE'S NOT what it used to be. As little as a decade ago, at the start of the 1970s, unlimited growth and unrivaled prosperity seemed to be North America's destiny. President Nixon and Prime Minister Trudeau presided over economies which were the envy of the world. Inflation was low, unemployment was low (so low that Canada's unemployment insurance fund was liberalized to dispose of its embarrassingly large surplus).

Energy problems had a different kind of urgency. America's nuclear industry, feeling it had gotten over its growing pains, was predicting more customers than it could handle. Canada's trade minister, fearing its oil in the ground would become valueless with the development of cheap alternatives, advocated selling off Canada's oil quickly while customers could still be found. Though peripheral events (like the Vietnam War and the War Measures Act) marred an otherwise perfect record, the majority of North Americans were feeling pretty smug about their position of prominence. Short of nuclear war, little loomed on the horizon to make us think that the future wouldn't merely be a bigger and better version of the present. It went without saying that energy growth would be behind the economic growth.

Our future today is not painted quite so rosily. High inflation and widespread unemployment have become structural features of North American life, and with the decline of the automobile industry predictions for our demise have become more dire. Depressions, labor strife, fuel shortages, conflagrations over Mideast oil supplies are all pointed to with despair.

Solutions seem beyond our capabilities. The once-growing nuclear industry has all but collapsed, and plans for energy self-sufficiency, like President Ford's trillion dollar Project Independence, or President Carter's War on Energy never got off the drawing boards. Plans that do get the go-ahead, such as Canada's tar sands scheme for energy self-sufficiency by 1990, seem to be products of some form of grim humor—should every plant be built on schedule, officials admit, Canada will be importing a greater percentage of its oil in 1990 than in 1980.

The old methods aren't working any more. Yet to get out of our present mess, we insist on intensifying the very methods that got us into it. The primary change in economic policy has been in the size of the subsidies handed to the energy industries. Super-depletion allowances have replaced the depletion allowances.

But if the future isn't going to be what it was supposed to be, neither is the present what it was meant to be. The normal free market mechanism by which we thought we were making our economic decisions was altered by intervention on a massive scale, spearheaded by the oil and automobile monopolies. This intervention did more than trade in our electric vehicles for gas guzzlers. It paved the way for the depopulation of cities and the rise of the modern suburb, creating an energy-dependent structure for both.

In an energy-short world, the situation is untenable. The suburb has thrown our energy balance out of kilter. We can maintain suburbs only at increasing expense to the cities that must support them. Unless we adopt a more flexible structure we will only be entrenching an unsustainable *status quo*, progressively limiting our options as we reach the limits of our fuel supplies. At this stage, all of society could face a complete collapse. But remove the still-existing monopoly interference and suburb and city can begin to reach some rough equilibrium, shifting away from suburban sprawl to realign themselves according to a sustainable structure. Whether we want them or not, the expressway and the low-density suburb have become a fact of life.

There can be no turning back of the clock, but there can be a return to the timeless principles that govern the way cities function and humans interrelate when left to their own devices. Reapplying those principles, we will be rejecting rigid, monopoly-made mistakes for flexible systems designed to adapt to nature's surprises.

These flexible systems, which were so familiar to us a half century ago, have all but been erased from human memory. Conditioned to fighting business rather than freeing it, we've suppressed our instincts, misplaced the necessary wherewithal of doing what comes naturally. In the case of the automotive industry—where GM emerged as the single biggest amnesiac—our wherewithal was lost in the 1920s. The organization that eliminated the competitive market in automobiles set its sights on control of the rest of transportation.

By 1934 General Motors' investment portfolio included Eastern Airlines, Western Airlines, TWA, and United Airlines. Douglas Aircraft, Bendix Aviation, and General Aviation, among other aircraft manufacturing giants, were already GM properties, and the United States Senate was concerned. This one company—already one of the largest in the world—would soon be in a position to clip the wings of the nation's aviation industry. By controlling the airline companies, GM could schedule flights at selective hours and provide air services at a cost and quality that would enhance the American's appreciation of the automobile. By controlling the airline manufacturers, GM could produce the precise technologies needed to serve these new needs of its airlines. The potential for massive interference with free trade practices—and consumer lifestyles—was clear. U.S. Senator Hugo Black co-sponsored a bill aimed at stopping GM, warning that if it wasn't enacted "our aviation industry is definitely headed for General Motors . . . control." The U.S. Senate passed the bill, and GM dropped its holdings in the four airlines.

The company was understandably upset. GM Chairman Alfred P. Sloan, Jr., explained it was only protecting its interests in the promotion of the automobile. "[We] got into aviation because we thought the . . . airplane would be an important competitor of the automobile . . . and we felt that we had to gain some protection by 'declaring ourselves in' the aviation industry." Sloan had some consolation, however. GM would be quite successful in its ongoing plan to buy out and dismantle the entire public transit system in the United States.

In doing so, it would be doing no more than fulfilling that contradictory credo of any competitive industry—wipe out the competition. Whether or not GM's future actions were premeditated, whether GM understood just how drastic a chain reaction it was setting off, is irrelevant. Good intentions or bad, the results would have been the same.

Let us give GM the benefit of the doubt, and credit the company with knowing exactly what it was doing when it tampered with transit systems that had been established long before GM was born.

By the turn of the century, every city in North America had discovered the trolley. Los Angeles alone boasted 3,000 vehicles, providing service on lines radiating 75 miles from the city's center. At its height in 1920, the streetcar carried close to 14 billion passengers a year. North America's love for the streetcar seemed unlimited, matched only by its infatuation for the automobile. Trolley companies were overextending themselves borrowing all they could to build new streetcar lines and obtain new franchises. Technological developments among the over 150 competing manufacturers of public transit vehicles flourished. This was the time of the steam- and electric-powered bus, of the double-deckers and the multiple-unit buses. All fashion of electric streetcars existed, and the high-speed electric interurban train made its debut.

Then in the Thirties the love affair suddenly chilled. Technological advance froze dead in its tracks. Maintenance and service problems began to plague the streetcar, irritating passengers to no end. More customers found themselves needing to transfer to other vehicles more often and waiting longer to do it. Public hostility to the arbitrary conduct of many streetcar companies led to their strict regulation by government. Strong, well-financed lobbies sprang up against the rail industries.

Almost overnight the Cinderella story of the streetcar ceased. Companies that had been able to do no wrong seemed unable to do anything right. A vehicle that had been part of everyone's future was becoming part of everyone's past. It was the first great collapse of a North American industry.

Economists of the day explained that the free market had spoken. Passengers knew what they liked, the theory went, and they didn't like the streetcar. The "invisible hand" of capitalism had sorted out the supply and demand of transportation systems and decided there was no room for the trolley. There *was* an "invisible hand" at work, but it was not the free market's. That invisible hand was General Motors'.

By the mid-1920s, the automobile market had already reached saturation. Those who wanted cars had them; new cars were being sold to old customers. General Motors looked elsewhere and decided to expand the manufacture of buses, which it found highly profitable. Blocking its ambition was the electric streetcar, which was faster, more comfortable, less polluting, cheaper, and more durable than the bus. This did not deter GM. Its executives simply sat down and worked out a marketing strategy that overcame all these obstacles.

According to the corporation's general counsel, General Motors "decided that the only way this new market for [city] buses could be created was to finance the conversion from streetcars to buses in some small cities." To test their theory, GM set up a company that bought trolley lines in Springfield, Ohio, and Kalamazoo and Saginaw, Michigan. Then the company dismantled the trolley lines and replaced them with GM buses. Its work done, General Motors got out. As GM's general counsel explained, in each case the company "successfully motorized the city, turned the management over to other interests and liquidated its investment."

Clouding this stunning success was the American Transit Association, a body comprised of streetcar interests not eager to witness their own destruction. Censured by the ATA for its actions, General Motors decided to dissolve its transit corporation. But GM soon saw this decision to be shortsighted. The bus was not going to easily make inroads into the major public transit markets like New York, Baltimore and southern California (an especially coveted market served by an extensive urban electric rail system and the world's largest interurban electric railroad network).

General Motors needed help, and found it in 1936 in the Omnibus Corporation, which it controlled. Together the two companies accomplished a feat that has since come to be recognized as the turning point in the electric railway industry—the conversion of the world's largest streetcar network (New York's) to GM buses in the space of eighteen months. Meanwhile, GM had joined with Standard Oil of California and the Firestone Tire Company to form a new transit corporation that picked up where GM's pilot project of a few years earlier had left off. Only this time the scheme was not limited to small towns. Transit companies, in debt and weakened by the Depression, were systematically bought out. Existing streetcars were replaced by new General Motors buses using Firestone tires and fueled by Standard Oil.

By March of 1949, when GM, Firestone, and Standard Oil were

convicted in Chicago Federal Court of having criminally conspired to replace electric trolleys with gasoline- or diesel-powered buses in violation of anti-trust laws, over one hundred electric transit systems in forty-five cities across America—New York, Philadelphia, St. Louis, Salt Lake City, Oakland, and Los Angeles among them—had been dismantled. The face and nature of the continent had been permanently altered. With transit under control, the car could now proceed unimpeded. The seeds of the future energy crisis had been sown. The road to an even more profound social change—the advent of the modern suburb—had been cleared and there was now no turning back. How profound and irreversible these changes would be could not have been clear in 1949. Only the magnitude of this interference in the free market system and the awesome abuse of economic power could have been adequately appraised.

General Motors was fined $5,000 and its treasurer, who mastermined the dismantling of the $100 million Los Angeles trolley system, was fined $1.

The court could see the laws had been broken, but who was the loser for it? The public? It only got what it wanted. The streetcar? It was on its way out anyway. The structure of the economy? If anything, GM had improved on it. To the court, a victimless crime had occurred, where events had unfolded as they should, only faster. To have levied greater punishment on some academic principle related to market structures seemed senseless.

While the streetcar systems were being secured, GM bought into rail and bus production and, along with Ford, Chrysler, and other highway interests, lobbied against rail transit. Through their dollars and by design GM would succeed not only in stunting all technological developments in transit but also in monopolizing it. Undaunted by the fear of facing another $5,000 fine, GM's early acquisition of Yellow Coach and Electro-Motive led to control of more than 85 percent of bus production by 1956.

The government filed suit again. Claiming that GM had unfairly manipulated its monopoly powers to maneuver scores of competitors from the market, the U.S. Attorney General spent the next nine years trying to get GM to court. Outmaneuvred and demoralized by GM's lawyers, the government finally dropped the case. It reluctantly accepted a consent decree whose principal provisions were drafted, at least in part, by GM's legal department.

Attorney General Robert F. Kennedy tried to take on General Motors as well, this time for monopolizing locomotive manufacturing. As GM produced 100 percent of passenger locomotives and 77 percent of all types of locomotives manufactured in the U.S., he thought he had a strong case. In 1961 Kennedy announced that a federal grand jury had indicted General Motors for the monopolization of the locomotive industry. The government was again claiming that GM had used its monopoly power to eliminate its competition.

After several years in the courts, the government again gave up. GM responded by eliminating Alco Products, the country's first manufacturer of diesel-electric locomotives. A decade after being sued for monopoly practices, GM maintained production of 100 percent of passenger locomotives while raising its share of all U.S. locomotives from 77 percent to 83 percent.

GM has succeeded in eliminating virtually all competing bus and train-locomotive producers. Having done so, it made sure to produce a relatively small number of bus and rail vehicles for domestic service. Road space was limited, and GM wanted to cover as much of it as possible with the private automobile.

Before 1923 the automotive industry was one of the most competitive on the continent. Auto producers were primarily assemblers of bodies and engines which were purchased from independent suppliers specializing in automotive parts. The interchangeability of component parts gave car producers great flexibility in selecting the best technologies at the best price for their products, while keeping the price of entry into the automotive field far from prohibitive. Henry Ford entered the market with only $28,000 cash in 1903 (about $250,000 in current dollars) and did so well that a few short years later he could refuse General Motors' offer to buy him out.

Others had less resolve than Ford. By 1909, GM controlled nine formerly independent auto assemblers, including Buick, Cadillac, Oldsmobile, and Pontiac. A dozen more were acquired by 1920, as GM went international to acquire the McLaughlin and set up a Canadian operation producing Buicks, Oldsmobiles, and Chevrolets at the McLaughlin factory in Oshawa, Ontario.

But eighty-eight U.S. assemblers were still left, and competition for the sale of 1.5 million passenger cars was keen. Alternative propulsion systems, in particular, were competing well. The Federal

Oil Conservation Board went on record as favoring electric cars, which it found superior to those with gasoline engines, while the steam car was reaching speeds of 190 m.p.h. (and meeting today's clean air standards).

General Motors (along with Ford and Chrysler) had a substantial investment in the internal combustion engine to protect, and good reason to fear the steam- and electric-car producers. They acted quickly and decisively, introducing marketing innovations that would not only eliminate the present threat from competitors but also assure no technological development could come along in the future to threaten their position. The industry was monopolized and the development of new propulsion systems halted.

In just three years—from 1923 to 1926—the number of automobile producers was more than halved from eighty-eight to forty-three. By 1935 ten remained. Today the number is four and still falling.

The first step was to deprive the automobile companies of access to independent sources of body and engine components. GM accelerated its rate of buying out parts manufacturers, and by 1937 could count over forty previously independent producers of key automotive components on its shelf. Ford made its own parts but Chrysler helped GM in taking over parts manufacturers.

The interminable process of buying out other parts manufacturers and other automobile assemblers eventually became an intolerable irritant to GM, and it sought a final solution to deal with those who would compete with it. GM began by introducing annual style changes in 1923, forcing other assemblers onto a treadmill they could not keep up with and freezing parts manufacturers out of more of their market.

The final curtain began to fall on the competition two years later, in 1925, with the end of the independent retailer. Car dealers, who had previously decided which automobiles to sell and service, were maneuvered out of handling competing cars. Dealers were bestowed with exclusive franchises, and there was little they could do to refuse this honor. The regular restyling of the automobile led to a decline in interchangeable replacement parts and a limit to the stockpiles a dealer could carry.

The consequences for assemblers were devastating. To stay in the game, an automaker now needed to establish nationwide dealer networks with specialized repair and maintenance capabilities. The small assembler was doomed, no longer able to distribute his cars through the thousands of previously independent dealers.

The revolution in the auto industry was complete, and the dominant regime—which came to be known as the Big Three—would enjoy half a century of uninterrupted rule. Through the acquisition of competitors, the control of component parts, annual style changes, and the exclusive-franchise dealer, the Big Three transformed a naturally competitive North American industry of well over one hundred competing automakers, thousands of competing parts manufacturers, and tens of thousands of independent car dealers into a single monopoly shared by three firms.

As in all unnatural monopolies, the inefficiencies are staggering. The technological advances of the pre-monopoly era atrophied into technological retardation. All innovation focused on style changes and planned obsolescence. To get more out of a car more had to be put in. Efficiency became a forgotten concept.

Between 1945 and 1972—the year before the OPEC crisis—the average gas mileage of the North American car actually fell by 10 percent as engines increased in power and compression ratios. The high compression engine, which goes no faster than the prewar engine, also brought us smog as a byproduct of its improved ability to accelerate us quickly (if not often) from 0 to 60 m.p.h. (Low-compression engines, which operate at lower temperatures, do not produce the kind of exhaust that generates smog.)

Before 1973 the car was consuming more fuel and more raw resources than any other commodity in the history of mankind. It was pushing environmental limits as the single biggest air polluter of all time, and safety limits as the instrument of 50,000 deaths per year. Half the land area of central cities were paved over to accommodate the car and North America was closing in on 5 million miles of road. For every mile of road there were about thirty cars, and for every mile of expressway forty-three acres of land were taken up.

North America became entirely dependent on the auto. The car was taken along on close to 90 percent of all trips for well over a trillion miles traveled per year, or the equivalent of about 5 million trips to the moon.

During the population explosion scare in the Sixties, when

concern focused on a North American gain of 25 million people, a larger gain of 34 million cars went unnoticed.

Along with the auto's record achievements in environmental inefficiency, social inefficiency, and resource inefficiency went the record for being the most fuel-inefficient mode of passenger transportation ever devised: neither the donkey-drawn cart nor the modern jet plane could touch the auto's inefficiency in passenger miles per unit of fuel consumed.

Not knowing how to deal with these massive misuses of resources, we began to regulate them. Literally dozens of government agencies at all levels became involved in the industry, redesigning vehicles and shifting production activity away from the continent's car assembly capitals to the government capitals. Bumpers, braking systems, seatbelts, headlamps, emission control equipment, tires, ignition systems, doors, power windows, head restraints, accelerator mechanisms, steering columns, door locks, seats, windshields, wheels, and fuel tanks now have the mark of a bureaucrat's design.

Not knowing how to deal with the automakers' massive misuse of resources, we added to the inefficiencies with additional inefficiencies of our own. The joint burden of the government and corporate bureaucracies will be more than the post-1973 world can bear.

Because the automobile is so enormously wasteful, massive and immediate efficiencies should have been available to counter gasoline shortages. But lineups at the pumps in 1979 followed the lineups of 1974, and it will not take another five years for the lineups to reappear. Gasoline rationing on a continent-wide scale is a virtual certainty, yet all that is being asked of the automobile is an improvement of two miles per gallon a year to 1983, a further one mile per gallon in 1984 and one-half mile per gallon in 1985. Industry's response has been to do what it does best—lobby hard against the new government regulations and lobby hard for subsidies to help meet them.

These asked-for improvements—which pale beside the energy efficiencies being reached in new houses, office buildings, and industrial processes, where energy reductions of 70 percent, 80 percent and 90 percent are becoming routine—will only result in the automobile fleets of 1985 consuming about as much energy as the automobile fleets of 1975. These goals are inadequate for our needs and will lead to irresistible pressure for a wide variety of vehicles to meet a wide variety of efficiency needs—cold-weather cars for the northern U.S. states and Canada, vehicles specifically designed for highway driving, short-range urban cars, cars meeting the needs of specific industries and specific geographic locations.

Monopolized industries—whose forte is applying one solution to all problems—are inherently incapable of responding to these needs. To avoid massive dislocations in the Eighties, the industry requires radical reorganization. The monopoly structure should be disassembled into its component parts by denying corporations human rights and allowing free enterprise to take over. Freeing the dealer from exclusive franchises and the small parts manufacturer from being forced to produce non-standardized parts will drop the price of entry into automotives fifteenfold, to about $80 million.

It will no longer be necessary to set up in-house production of specialized body and engine components or to produce annual style changes or to maintain a national dealer network. As a result literally thousands of entrepreneurs and enterprises will again have access to the field, unleashing generation after generation of technologically improved vehicles.

Canada (in much the same way Brazil developed cars to run entirely on alcohol from its sugar cane plantations) will develop rust-free, cold-weather cars that run on liquefied forest fuels, or, in regions like Quebec that have vast oversupplies of waterpower, on electricity. Corn belt states will fuel their vehicles with liquid fuels from animal fodder; sun belt states will develop solar-powered cars to convert sunlight directly into electrical power that could be used or stored, as needed. The steam car will make a rapid resurgence in thousands of rural areas.

Such diversity would mean automobile assembly plants spread out across the continent, each producing enough to meet the needs of its specialized markets. Production runs would vary from 25,000 vehicles per year—or even fewer when highly specialized needs were required—to production runs of 200,000 or more for high-volume needs such as cold- or warm-weather machines. Runs on this scale would surprise no one in the automotive field. Because of diseconomies of scale, auto production today already takes place in about fifty separate locations in the U.S. and Canada, each averaging some 200,000 cars per year. European manufacturers, which have little trouble competing in North America, typically produce fewer cars per plant due to higher standards of workmanship and fuel efficiency.

In West Germany, France, and Japan, whose combined population and total car production approximates North America's, there are over 300 different models available from twenty competing producers of passenger vehicles, most of which have avoided GM's pattern of production. Companies like Porsche and Suzuki buy, rather than make, their parts, providing economies of scale for many parts manufacturers who supply the entire industry. European and Japanese automakers generally distribute their products through non-exclusive, independent dealers. They achieve high levels of efficiency by using standardized, interchangeable components and by avoiding annual style changes. Production runs vary from Porsche's 15,000 to 200,000-plus for lower-cost passenger cars. Even Japanese firms like Mazda, however, were innovating a little over a decade ago with production runs under 130,000 cars per year.

Though the European and Japanese car market is hardly a paragon of the free market (domestic monopolies exist in many European countries), the need to export a substantial part of production (unlike here, where 90 percent of production stays at home) has led to a considerable measure of competition.

The absence of the monopoly structure has resulted in a technological excellence unmatched by the large North American producers. While the Big Three can point to no innovation during its last forty years more impressive than the concealed windshield wiper, it was left to the foreign manufacturers to introduce low-pollution cars, long-life cars, and standard safety equipment such as the safety belt, crash-absorption bumper, interior crash padding, and collapsible steering column. While North American producers offered only one propulsion system—the gasoline engine—Mercedes offered the diesel, Honda the stratified charge, Mazda the rotary, and Daihatsn the electric. Disc brakes were standard equipment in Europe in 1955 and the radial tire has been popular for almost thirty years. North American innovation, when it did occur, came almost entirely from the research and development of small firms or independent automotive suppliers. It was Dusenberg who pioneered four-wheel brakes (1920), Nash who brought us the quiet motor (1922), Reo who developed completely automatic transmission (1934). Power brakes, nonslip differentials, and transistorized ignition systems were also left to the ingenuity of the entrepreneur.

In today's resource-short world, we need such innovation more than ever before. Unleashing North American ingenuity by reorganizing General Motors, Ford, and Chrysler into the dozens of automotive assembly plants and scores of parts production facilities they had previously acquired—in other words, restoring free enterprise—would not only result in immediate benefits for society but help the captive companies and their shareholders.

With few exceptions, previously freed companies have done exceedingly well. TWA, Eastern, Bendix and North-American Rockwell have all taken off since their release from GM. In 1936, TWA's year of separation from GM, its operating revenues were $6.2 million. By 1961 they had soared to $1.1 billion. Eastern grew from $3.8 million in 1938 to $649 million in 1969. Bendix's sales of $162 million in '48 leaped to $1.6 billion by 1971, and North-American Rockwell, the aerospace outfit, turned its $11.7 million of 1948 into $2.2 billion in 1971.

Hertz, acquired by GM in 1925 when it feared the rent-a-car business might encourage consumers to use cars only when needed, was deliberately suppressed until 1953. By then GM had succeeded in establishing the rental car as a rent-away-from-home institution. When GM no longer feared Hertz, GM dropped it. In Hertz's last five years under GM control, its revenues increased modestly, from $11 million to $17.9 million. In Hertz's first five years out of GM's grasp, revenues almost quadrupled, from $20 million to $79 million. Under GM for twenty-eight years, Hertz increased the number of cities it served by three, from twenty-seven to thirty. In just five years after GM, Hertz cars were marketed in an additional 148 cities, for a total of 178.

A single rental car can displace the need for several passenger cars. One trolley coach or bus can eliminate the need for thirty-five cars. One streetcar, subway, or rapid transit vehicle can supplant fifty cars. An intercity railroad or train can replace 1,000 cars or a fleet of 150 cargo-laden trucks.

While the car is an indispensable form of transportation, it is not indispensable for 90 percent of all trips and it is economically and socially destructive in the present perversion of its use. To make the auto efficient, it must not only be built efficiently but used efficiently in a transportation system without suspect alternatives. For the private vehicle to be legitimized, the public transit systems that compete with it in cities, outside cities, and between cities must be rehabilitated as well.

Crippling the internal combustion engine's intercity competition was not easy. Electric intercity rail—whether for passengers or nonhuman freight—knows no rival, certainly not the diesel-driven machines GM promoted in their stead. Diesel engines, while they get better mileage and longer engine life than gasoline engines, their petroleum sisters, cannot compete in any capacity with the more fuel efficient, more durable electric. As any scientist or engineer knows, liquid fuels are by their nature not as well suited to turning motors as electricity. That the continent could overlook this elementary bit of information for decades can only be credited to the raw persuasion of one of the most powerful lobbies ever assembled in peacetime.

# 10
## *On Track*

TO PROMOTE THE urban use of its vehicles, GM assumed control over the mass transit systems in cities, replaced streetcars with buses wherever necessary and then replaced buses with cars wherever possible. To promote the highway use of its vehicles, GM needed to control another railed rival—the railroad. This meant converting the electric rail system to diesel locomotive. GM buses and GM trucks had an otherwise limited future.

Part of the job had already been accomplished as a side benefit of eliminating the streetcar and with it, much of the electric interurban system. With the destruction of Southern California's network went the destruction of the Pacific Electric Railway, which not only annually transported 80 million passengers through the region's 56 incorporated cities but also ranked as the third largest freight railroad in California, interchanging cargo with the Southern Pacific, the Union Pacific, and the Santa Fe.

In Southern California and elsewhere, merchants who once benefited from local deliveries made on streetcar tracks found themselves making other arrangements once the tracks were torn out.

The degradation of the railway system required entirely different approaches from those GM had been accustomed to. Locomotives did not lend themselves to annual model changes, and new competition for GM's diesels could not be kept out of the market through a local dealer network.

Compounding these problems was the diesel locomotive itself.

Though it could outperform steam locomotives, it lasted half as long, did one-third the work, and cost three times more than the electric locomotive. Diesels were widely considered sluggish, noisy and generally less attractive to passengers than the electric train—so quick, quiet, and clean of exhaust. Freight also rolled less efficiently on the less powerful diesel.

In 1930, when GM bought out and merged the two largest diesel manufacturers on the continent to become the single largest locomotive manufacturer, it had no effective marketing strategy to eliminate the electric locomotive. In 1935, when it developed one, the trend in trains was turning to the electric. By then, two of America's major railroads had electrified substantial parts of their systems with more eyeing the overhead wires. The Pennsylvania inaugurated freight and passenger service on its New York-Washington run while the New York, New Haven & Hartford now operated 500 miles of electrified track in New England.

General Motors reacted to these provocations in November 1935, by beginning a boycott of all railroads whose locomotives were not stamped "GM-made." Its traffic division began routing freight onto railways which agreed to scrap their electric and steam locomotives for GM diesels. Any railway that thought of refusing this offer of "reciprocity" risked losing the business of the largest shipper of freight on the continent.

Sometimes high-level intervention was required. In November 1948, GM's chairman, Alfred Sloan, personally offered to locate a GM warehouse on the tracks of the Baltimore and Ohio Railroad if B & O's president, Roy B. White, reciprocated by switching to GM diesels. White's reply came later that month: "Here is your Christmas present . . . we will purchase 300 diesel locomotives . . . we now expect a New Year's gift from you . . . locate your warehouse near our tracks."

General Motors' legal department, chagrined by upper management meddling, discouraged personal pressure. As an interoffice memo from its anti-trust attorneys stated: "GM could, in all probability, have successfully capitalized on the railroad's sensitivity to reciprocity by frequently reminding them of GM's considerable traffic, and could have done so without ever interfering substantially with the economic routing of traffic."

Some companies, like the Gulf, Mobile & Ohio Railroad, tried to cut corners, converting only partly to diesels. GM informed it that some traffic would be rerouted because other railroads had purchased more GM diesels. As GM saw it, it would be unfair to cut Gulf off entirely.

The New Haven railroad, one of the first to electrify, was also one of the last to switch. In 1956 it caved in to GM pressure and agreed to scrap all its electric passenger and freight locomotives in favor of GM's diesel. During its entire fifty-year history of electrified operation the New Haven had never failed to make an operating profit. In its last year before dieselization the New Haven earned $5.7 million carrying 45 million passengers and 814,000 carloads of freight. Three years after it began to switch to diesel the New Haven lost $9.2 million carrying 10 million fewer passengers and 130,000 fewer carloads of freight. Two more years, and the New Haven was declared bankrupt.

When the U.S. government investigated the collapse of the New Haven it found the conduct of the railroad inexplicable: disregarding the overwhelming advantages of the electric locomotive and against the advice of its own independent engineering consultants, New Haven chose to accept GM's claims about the superiority of the diesel—claims the government investigation characterized as "fabulous," "erroneous," "inflated," and "manifestly absurd." General Motors was censured for contributing to the railroad's ruin and the trustees of the New Haven were advised to undertake a study of the economic feasibility of re-electrifying the New Haven's main line.

The story of the New Haven was repeated in numerous locations across North America. New Haven passengers, used to rapid rail service, refused to be serviced on the slower diesels. Being less powerful, the diesels also did a poor job of delivering freight. These factors, coupled with the diesel's higher operating and depreciation costs, led to less revenue which led to cutbacks in maintenance and service, which led to even fewer customers—a downward spiral from which the New Haven would not recover.

By eliminating the electric railways, General Motors was also eliminating its locomotive competition. In 1935, when GM decided to rid the rails of electricity, electric locomotives outnumbered the diesels by seven to one. In 1973, the diesel outnumbered the electric 100 to one (with most of the diesels being manufactured by General Motors).

GM had managed to eliminate all but one of its competitors. The casualties included Westinghouse, one of the pioneers of the electric locomotive, which scrapped its operations in 1954. Baldwin-Lima-Hamilton, among the first railway builders to enter the field, left in 1956. Among the last to try to enter the field was Fairbanks-Morse, born 1944 and broke by 1958. American Locomotive, too aggressive to be forced out, was bought out in 1969 by a GM subsidiary which immediately discontinued locomotive production.

Along with the elimination of the electric locomotive went the deterioration of rail service and the domination of highway vehicles. Since 1935, the railroad has steadily been supplanted by the bus, the truck, and the car. In 1939, the railroad carried 500,000 passengers and three-quarters of all freight. By the time of the 1973 oil crisis, it had been deserted by half its passengers and three-quarters of its freight. Rail passengers went to the bus and rail freight went to the truck. But the revenues were all going into the same coffers.

GM, Ford, and Chrysler took over trucking at the same time and in the same manner they took over the automobile. Once, the trucking industry was as fiercely competitive as the auto industry with over one hundred producers of a wide variety of vehicles. Today, the Big Three account for 80 percent of production—the only others left with more than 1 percent of the market are American Motors Corporation and International Harvester. For good measure, and to benefit directly from the railways' loss, GM also bought into two of the largest trucking firms. Through them, GM established over one hundred freight terminals across the United States. By 1973, about 80 percent of all freight revenues went to the trucking industry.

At one time, entry into the trucking field was relatively easy. GM took over the parts manufacturers (as it was doing with cars) and introduced the nationwide network of franchised dealers for trucks. Since the 1920s not one firm has entered truck manufacturing and survived. Technological advances have been left to either the few remaining independent truck producers and suppliers (anti-skid braking and semi-automatic transmissions) or to the Europeans and Japanese (electric-powered trucks and vans).

Competing bus manufacturers were similarly squeezed out. Over fifty firms have left the field since 1925, and with them technologies like the double-decker bus, the "bendable" bus and the steam-driven bus—technologies North America will increasingly be forced to rediscover after their neglect of a half-century and more. (Lear Motors, makers of the Lear Jet, has developed a quiet, low-polluting steam-turbine bus that has yet to crack our GM-dominated bus market.)

Unlike the auto industry, where GM chose to let only its franchised dealers sell cars to the consumer, or the truck industry, where GM chose to let the truckers deliver the goods to their clients, or the train industry, where it did not assume direct control over the railways, GM felt it had a more prominent part to play in promoting the bus.

The year after General Motors became the continent's biggest bus manufacturer by buying out Yellow Coach, it helped form the company which would in 1928 boldly announce its intention of converting commuter rail systems to intercity bus service. This company, which GM and its co-creators named the Greyhound Corporation, agreed to buy virtually all its buses from GM. General Motors reciprocated by refusing to sell any buses to Greyhound's intercity competition—effectively eliminating all other bus operators. By 1952 both GM and Greyhound had their monopolies. By 1973 GM's only competitor, Motor Coach Industries (established by Greyhound under court orders in 1962 after an antitrust suit) was completely dependent on GM for its parts. Trailways, Greyhounds's only competitor, was boycotted by GM and forced to buy European buses, whose prices became prohibitive after the GM lobby inspired high duties. To the railways' surprise, Greyhound made good on its 1928 prediction. Under pressure, the railroads found they had no choice but to share their business with Greyhound. Pennsylvania Greyhound Lines took over substantial traffic from the Pennsylvania Railroad, Pacific Greyhound Lines from the Southern Pacific Railroad, Southwestern Greyhound Lines from the St. Louis Southwestern Railroad, and Greyhound Lines of Canada from the Canadian National and Canadian Pacific Railroads. In some cases, including portions of the world-famous, high-speed Chicago North Shore Line, Greyhound took over and transformed electric interurban commuter service to its own needs. By 1950,

Greyhound carried about half as many intercity passengers as all the railroads of North America combined.

Greyhound's 1928 prediction is all the more impressive when it is realized that the road system then was not designed to provide the kind of comfort Greyhound would need to lure passengers away from their trains. Today's equivalent of literally hundreds of billions of dollars would need to be invested in upgrading and expanding the road network—a program that dwarfs the American investment needed to place a man on the moon and eclipses the effort needed to finance the Vietnam War.

Greyhound knew that GM's plans were proceeding for a formal alliance of all the highway interests, including the American Automobile Association (which claimed to speak on behalf of all motorists), the Motor Vehicles Manufacturers Association (which did speak for the automobile and truck companies), the American Truckers Association (which represented the trucking interests), the Rubber Manufacturers Association (which spoke for the tire companies), and the American Petroleum Institute (which mobilized the considerable clout of the oil industry). This lobby, which went public in 1932 as the National Highway Users Conference, became the most sweepingly influential re-educator of congressmen and state legislators the United States has ever seen, acting where "the membership may be badly informed or where a considerable part of it may yield to the influence of selfish interests."

In the four decades to the oil crisis, the highway lobby—by building up some 2,800 lobbying groups with an annual lobbying budget of $500 million—had moved mountains of money, earth, and congressmen. Forty-four of the fifty state legislatures had agreed to pass specific legislation that would make state and local gasoline taxes legally unavailable for any purpose other than construction of highways.

At the federal level, the highway lobby was the chief architect of the world's largest roadbuilding effort—the $70 billion, 42,500-mile Interstate Highway System. The campaign needed to have that sum allocated took years; it involved winning over federal politicians and the federal bureaucracy, along with associated academics and engineers; it required the granting of millions of dollars to highway research, contributions to congressional campaigns, and the placement of lobby members in key administrative posts.

So well had the lobby done its preparation that during more than two years of congressional hearings only one witness among the many academics, engineers, and other experts raised the issue of what effect so large a highway project would have on the railways.

The effect was predictable. Between the passage of the Interstate Highway Act in 1956, and 1970, the federal government alone spent $70 billion for highways. During the same period, when strong measures were clearly needed to preserve the railroads—only $795 million, or just over 1 percent, was spent on rail transit.

The railroad and transit lobbies, outspent by more than ten to one and confused by General Motors' strong voice from within their own ranks (as locomotive manufacturers, bus manufacturers, and transit company owners), were no match for the highway interests. Even politicians with the stature of Senators Kennedy, Muskie, and Weicker, or Mayors Alioto (San Francisco), White (Boston), and Daley (Chicago), were powerless in repeated efforts to shift anything but token amounts of gas-tax monies from highways to rail. While auto registrations doubled over the two decades to 1973, transit declined by half.

North of the border, the legislators have accepted the same formula with the same effect—money is always available in large amounts for freeways but subways and railroads have to compete with hospitals and schools for their funding. The advisability of building roads instead of railroads was not questioned.

The highway lobby has had few setbacks in Canada, in most provinces raising billions more for roads than gas and motor vehicle taxes brought in, and succeeding in many cases to have the railroad removed from the center of cities. In 1950 its views were well represented in Jacques Creber's "Report on the Plan for the National Capital," which recommended the removal of eleven individual railway lines from the twin cities of Hull and Ottawa. Among the reasons cited was the standby argument that trains contribute to traffic jams.

Railroad crossings are often pointed to as a cause of congestion, when all that is really being pointed out is a pro-car bias. From the car's point of view, level crossings are an annoying impediment, and elevated crossings just another expense incurred by the railroad. This perspective assumes the car as the norm and the railroad as the interloper. Yet the railroad came first, the expense of elevated railroad crossings is made to suit the car, the railway is far more efficient than the car, and from an overall point of view, where it is transportation that is optimized and not any particular kind of transportation vehicle, the best way to relieve the congestion brought on by the car is to increase use of mass transit.

Following Creber's report, agreement was reached between the National Capital Commission, the Canadian National, and the Canadian Pacific Railways to remove all major railway installations and tracks from the central area of Canada's capital region. Expressways were built in their place, with city planners proudly pointing to the six-lane Ottawa Queensway as evidence of the excellent use that can be made of abandoned railroad rights of way.

While one part of the highway lobby (the prosperous Canadian Trucking Association) argues forcibly against freight subsidies to the railroads that, it says, would wreck trucking because of the low fares clients would be offered, another part of the highway lobby (Greyhound Lines of Canada) has played a key role in keeping passenger fares on trains artificially high, discouraging intercity rail use.

Until just before the energy crisis of 1973, General Motors and its lobby (which continued to be organized and directed by its own personnel) were gathering momentum for a final assault on what little remained of public transit and the rail industries. To replace the railroad as a freight carrier GM was developing 750-horsepower diesel engines to haul "truck-trains" at high speeds along its Interstate Highway System. Triple-trailer truck trains, which occupy 100 feet of highway, had already become commonplace in parts of California, Nevada, and Oregon.

Instead of subways and commuter rail systems, the GM lobby was pushing "bus-trains" of up to 1,400 units which would operate on exclusive busways outside cities and in bus tunnels under downtown areas.

These bus-trains may have been slated to eliminate not only passenger rail but also the one-unit bus which, having done its part to outmode the passenger train, was becoming more and more an impediment to additional car sales. Electric locomotives have a thirty-year life, diesel locomotives last fifteen years, heavy-duty trucks seven and cars five. One train displaces 1,000 cars or 150 trucks. Every time General Motors could replace one of its passenger diesel locomotives with cars its revenues went up by a factor of thirty-seven. When a freight locomotive was replaced with trucks, revenues went up twenty-seven times. (Had GM been producing electric locomotives, the factors would have been seventy for cars and fifty for trucks.)

Replacing locomotives with buses instead of cars produces far less revenue. A profit-motivated company would have an incentive to eliminate the bus. GM was such a company. From 1952 (the year GM achieved monopoly control over bus production) to 1972 (the year before the oil crisis) bus sales fell by about 60 percent and bus ridership by almost half. GM had already begun converting its bus plants for a market with more mammoth potential—the motor home.

Until the oil crisis, the motor-home industry was the fastest growing sector of the automotive industry. Trailer parks were already well established. Luxury homes-on-wheels were being equipped with self-contained dining, sleeping and bathing facilities in a wide variety of interior designs. Sales to retirees—once thought of as the extent of the market—were viewed as the tip of the iceberg, with the motor-home manufacturer's realization of the "big growth market among younger people [who are] willing to spend for the kind of life they want to lead."

Airlines were offering fly-drive packages with the "drive" part being behind the wheel of a motor home while GM had established a nationwide motor-home dealership network that offered emergency around-the-clock "sudden service" for traveling families who needed a plumber to unstop their toilets or an electrician to repair lighting fixtures.

In the same way the conventional train, streetcar, and bus were being outmoded, the conventional non-mobile home may have been on its way out. GM was predicting that soon everyone would own "a family of cars" for every conceivable travel activity including "small" cars for trips, recreational vehicles for leisure, and motor homes for mobile living.

In September 1972 General Motors' chairman R. C. Gerstenberg expressed optimism that "expanding markets all around the world give us the historic opportunity to put the whole world on wheels."

The 1973 OPEC oil crisis put the brakes on all of GM's plans. Scarcity of fuel supplies forced GM to shut down motor-home plants and governments to refinance transportation systems.

The six bankrupt railways of the U.S. Northeast and the two poor performers of Canada were resuscitated (with subsidies), renamed (Amtrack south and VIA Rail north of the border) and relieved from their plight in the private sector (in the U.S., the railways were nationalized; in Canada the CNR was already government owned). Multi-billion dollar commitments were also made in public transit in efforts to revive the rail—Washington, Atlanta, Calgary, Boston, Edmonton, Baltimore, Buffalo, Cincinnati, Vancouver, Santa Clara County, Pittsburgh, and others were injected with large amounts of money.

These systems are all destined to deteriorate. The changes have been cosmetic, amounting to little more than a fresh coat of paint (in the case of the railroads) and resulting in public transit systems that exist by the grace of government, carrying a bigger financial burden than politicians in the past have been willing to maintain.

Until rail can be made to pay its own way, it will be ever vulnerable: aside from the military, few subsidized services last long without being degraded by governments as soon as it becomes fashionable to be fiscally responsible. The pressure may come from the public, through such measures as Proposition 13, or from large lobbies that don't like efficient public transit.

Rather than remodel the railroad, we need to rejuvenate it by restructuring the highway industries that so remarkably sapped our far-more-efficient rail systems decades ago.

The restructuring of the trucking industry—breaking it down into its component parts—will be largely accomplished as a natural consequence of the restructuring of the Big Three's automotive assembly plants (which for reasons of efficiency are already independent plants in all but ownership). Three-quarters of Big Three truck plants assemble automobiles as well, and these could become independent car/truck plants. On the assumption the Big Three know how to maximize profits, the remaining one-quarter would remain all-truck assemblers.

Disassembling bus and locomotive manufacturing would do much more than restore competition and incite technological revolutions: it would free the transit lobby of GM influence and break up the

GM-dominated highway lobby, letting its diverse interests express themselves. With bus operators being able to compete for part of the auto market, and car/truck assemblers finding that their interests didn't always coincide with those of trucks or cars, and manufacturers of both diesel and electric vehicles not eager to see either market destroyed, none would any longer have to pretend that what was good for GM was good for them.

Though still potent, the size of the lobby would be reduced to about the proportions of the transit lobby, providing much needed political balance and the incentive to compete on product quality instead of political inequality. This can best be seen in Europe and Japan, where the auto industries are competitive with each other; and the rail and bus industries are competitive with the auto. Unlike North America, these countries are largely free of highway trust funds and other devices designed to promote one form of transportation at the expense of another; they spend roughly half their transportation budgets on rail and half on highways.

The result is two good systems instead of one, accompanied by the option to choose on the basis of personal preference rather than by default. Personal preference will, on the evidence of much of the world, generally correspond with transportation efficiency.

While over 90 percent of intercity travel in North America is by car the figure is less than 10 percent in Japan. While most freight in North America is trucked, most European freight travels the rails.

Rail has advantages over road in areas of high density, such as Japan, and low density, such as Sweden. Were Sweden to be flipped on its back, its 1,000-mile-long, 200-mile-wide shape would roughly resemble Canada's population distribution, which hugs the Canada-U.S. border. Yet unlike Canada, over 60 percent of Sweden's 7,000 miles of track is electrified and 93 percent of passenger and freight traffic is hauled electrically.

Competitive transportation markets in Europe and Japan have produced transport systems no car can match. Passengers on Japan's "Bullet" trains travel in luxurious comfort at speeds of 150 miles per hour, arriving at their destinations with such consistent punctuality that watches are reset whenever the train arrives late.

New generations of Japanese trains are being designed to move passengers at speeds of 200 and 300 miles per hour, but North America need not wait for them to come on stream—train service averaging 125 miles per hour—the precise service which presently exists in Britain and Japan—would be adequate to destroy present-day notions of convenience and convince consumers to abandon both of the poor-efficiency passenger vehicles we have: the automobile and the airplane.

Unable to compete with the train for travel between cities, the car will turn to functions for which it is highly efficient: luxury uses such as camping and excursion trips, and practical uses like transportation to the many low-traffic areas that cannot support mass transit systems.In these, the car will prove unassailable and absolutely indispensable. Less secure will be the plane, which will find its place in transportation increasingly as an over-the-water and over-the-hill convenience.

A train which averages 125 miles per hour could cross the continent in one day, the new generation of Japanese trains would make the trans-continental trip overnight, and the underground train system suggested by the Rand think-tank would take half an hour from New York to Los Angeles.

But even at present-day speeds of 125 miles per hour, major North American airports would find their total traffic cut by one-third or more as the highest-passenger air corridors on the continent were grounded. Most of the travel in the Northeastern United States and Canada's Quebec City to Windsor corridor would now take less time downtown to downtown by train than by plane, sparing businessmen the necessity of holding meetings at airport expressway hotels and society the need to expropriate more land for more airports.

Development along airport strips would shift back to downtown cores, which, as aerial photos of most U.S. cities show, best resemble London after the blitz.

But to revitalize North American cities, urban transportation systems will need a new set of accountants. Even more than a revamping of the system we presently have in place, we need to simplify the arithmetic we've been using to understand the relative costs of our various transportation options.

Had we known the total costs of the systems we adopted, this profit-motivated society might have acted somewhat differently.

# 11

## *What the Traffic Will Bear*

PUBLIC TRANSIT DOES not pay. On this there is no dispute. There isn't a major city on the continent whose transit system didn't lose money last year, won't lose money this year, and isn't expected to lose money next year. Understand the motorist's chagrin, then, at having to support a system he does not use, whose vehicles deliberately stop at green lights in front of him to pick up precisely those people whose tickets he's subsidizing. These people then go on to act noble for using public transit while making him feel guilty for using the oil companies' overpriced gasoline which the public-transit-promoting government wants to price even higher.

Under these circumstances, the motorist should have every right to be upset. Only, these aren't the circumstances.

While it is universally understood that the public transit systems are subsidized, it is not generally known that the subsidy to the automobile far exceeds the subsidies paid to transit. It is the motorist who has been getting handouts from the rest of society. Most of the direct costs of roads and highways, the highway lobby tells us, should be paid for by the users through taxes and tolls the non-users do not contribute to. It would be undemocratic, the lobby explains, to foist costs on those who don't benefit from them. But through a perversion of the tax process, this is exactly what has happened. The North American's wealth has been appropriated for the welfare of the North American motorist.

The state owns minerals in the ground the same way it owns the lakes and rivers. They are the common property of all citizens. For the same reasons Floridians would want a share of the profits if Alaska were sold back to the Russians, the revenue from the oil lands is the common property of all citizens.

When the state puts its oil lands on the auction block, companies bid for the right to produce and sell the oil they find. In a competitive market bidding for property that's especially valuable, the oil companies have to bid higher for it, giving the state higher royalties. In a monopolized market, the low bid wins every time.

Royalties, as it turns out, make up only about one-eighth of the cost of gasoline. Most of the costs come from production expenses, oil company profits and government sales taxes on the final cost of gasoline. These sales taxes would normally not amount to much—they are generally meant as taxes on production, taxes on the use to which that product is put. But in the case of gasoline the sales taxes comprise most of the costs. They are not true sales taxes but supplements for royalties that were kept artificially low by the oil industry—a monopoly over a century old.

Because the government did not get its due from the sale of mineral rights to the oil companies, in compensation the government collected taxes after production. This allowed the government to pocket the same amount of total revenue, but, instead of considering the compensation as royalties, it was treated as taxes on production. These taxes were quickly claimed by the highway lobby for the purpose of laying asphalt—giving motorists a mostly free ride on an ever-expanding road system. The money that by rights belonged to every citizen went to subsidize the users of gas-powered vehicles.

The free ride now applies to oil that is imported from abroad too. Every barrel of oil that Gulf or Shell or Sunoco buys from the OPEC countries entitles them to from one-third to one-half the cost of the barrel back—with government providing the rebates out of taxpayer money. So huge are these subsidies that they are a chief contributor to the federal debt of both U.S. and Canadian governments. European motorists, not used to these helping handouts from non-motorists, typically pay two to three times as much as we do for gasoline.

The free ride for the automobile extends beyond the roads and fuel: it includes the costs of police patrols and traffic control, the costs of providing accident victims with hospital beds and medical attention, the costs of untaxed land unproductively used for roads and parking lots, environmental costs such as noise pollution (which lowers property values while raising tensions), and the costs of health and property damage resulting from road salt and exhaust-pipe fumes.

These are not insignificant costs. Over $10 billion per year, for example, is lost to the North American economy because of automobile pollution's attacks on paint and metals. The $10 billion figure is rising now that the toll from acid rain—the loss of tens of thousands of our lakes—is about to be factored in. Most damage occurs in urban areas—about 70 million tons of toxic pollutants fall in high concentrations over valuable houses and machinery.

Human life, not easily valued in dollars and cents, must be reckoned by body counts. Los Angeles operates the busiest automotive morgues, chalking up over 500 deaths per year from ailments attributable to motor-vehicle-generated smog.

Public transit of course shares in some of the benefits and blame. These deaths do not come only from carbon monoxide and hydrocarbon emissions from cars. Diesel trucks and buses also produce nitrogen oxides which, when inhaled, combine with moisture in the lungs to form nitric acid which permanently damages lung tissues and accelerates death by slowly destroying the body's ability to resist heart and lung diseases.

As well, 1 percent of highway use is by public transportation, so 1 percent of the costs associated with highways can be counted as furthering social goals. But the vast majority of these costs—all paid for by car driver and non-driver alike—benefit only the car driver. Unknown to the pedestrian who curses the car, he has played a prominent role in supporting its use.

Added to these traditional fuel, environmental, and social subsidies is a new generation of subsidy, known by a variety of euphemisms in the trade, but best understood by the public as bribes to corporations. Unlike past practices, where public money went directly to the public good (or at least the motorist's good) to keep prices of gasoline low or the quality of roads high, public money now goes to the corporate good to enable business to wreak its benefits on the public.

The greatest good, governments have been told, is wrought by the automotive industry. As with any other good, governments must pay for it.

Pennsylvania was first to pay, luring Volkswagen away from building North America's first VW plant in Ohio. For its $250-million investment, Volkswagen received a combined package of subsidized loans, tax holidays, and capital construction worth over $70 million. This lesson was not lost on Ohio, which made sure to outbid Michigan in 1977 for a large Ford transmission plant. As the automotive industry comes to recognize this large, new untapped government money market, however, Ohio may have difficulty staying in the bidding. Ohio had the chance to pay Ford to expand its engine plant in Lima, but in making its offer, Ohio felt that local ties in the town would be worth something to Ford. Ford, Ohio, decided, would take its money and stay in Lima.

Ford had a better idea and moved north to Ontario, a place still smarting from its loss of a GM diesel plant to Quebec. In Canada, Ford was made welcome—the federal government gave Ford $40 million outright and the provincial government gave it $28 million outright (while also promising to spend another $35 million to build an expressway to the plant and provide other sundry services). Ontario was so pleased with its ability to sell the province to industry that it set up a $200 million fund to maintain a competitive advantage in future. (Ford, it turned out, took its money and ran when the new plant became inadvisable; the company earned $68 million for good intentions.)

The most important subsidy to the automobile, however, remains the benefits that the car lobby brings it. In 1971 the lobby persuaded President Nixon to pass legislation dubbed the "auto industry relief act," to spare the Big Three from having to produce small cars. As Henry Ford II explained it that year, "minicars mean miniprofits," and without a surcharge on the small imports, the Big Three would have had no choice but to take smaller profits or face unbearably stiff competition from the foreign makes. President Nixon outdid his auto relief bill a few months later, when it looked as if the surcharges on foreign cars might not be enough to keep out Japanese vehicles. He persuaded the Japanese government to impose voluntary quotas on their cars the next year.

Since 1949 the auto industry had resisted pressure from the United Auto Workers and others to build small cars, despite marketing surveys at the time showing six out of ten Americans wanted smaller cars of the type the Big Three was manufacturing in its foreign plants.

By 1980 the cumulative subsidies to the North American automobile industry—over and above subsidies normally available to business—totalled hundreds of billions of dollars.

In contrast to this half-century-long history of direct and indirect handouts to the private vehicle, subsidies to public transit have been on a relatively small scale and are only a recent phenomenon. Numerous public transit companies on the continent were making ends meet until the Sixties, and it wasn't until the Seventies that large deficits started to occur. This despite the onslaught of the auto manufacturers, the scrapping of streetcar systems and trolley buses for diesel buses, and a public relations campaign that succeeded in convincing most North Americans that public transit was for the underprivileged.

The systems that fared best were the ones furthest from General Motors' grasp—those in Canada. And the best system in Canada is the one that is also the most diverse, having commuter links by bus, rail, and ferry in a city that has kept its train station in the downtown core. That city is Toronto and its municipal transportation system uses diesel buses, trolley buses, subways and . . . the streetcar.

Toronto's transit system (the TTC) fought off the automobile so well that—until 1970—it was self-sufficient, needing no subsidies to meet its expenses. Even into the Sixties, it was still seen as a rich relative that could be exploited, paying taxes and getting nicked with additional charges at every turn.

The privilege of bringing customers to the Canadian National Exhibition—Toronto's summer carnival—cost the TTC over $11,000 a year. Because the TTC decided to maintain good service in winter by clearing its tracks of snow, the city charged it an additional $50,000 a year to cart away the snow ploughed by TTC vehicles. Despite these charges the TTC managed to keep several million dollars in the bank.

Toronto was General Motors' most dismal failure, but GM can hardly be faulted. As with most Canadian cities, Toronto's transportation system was publicly run by the time GM launched its streetcar-scrapping program in 1932.

Unable to buy transit systems from governments, GM concentrated on the next best thing. Long before the Washington Society of Professional Engineers would declare that "dieselization results in certain bankruptcy of urban transit systems," GM knew that its hopes for the car hinged on the shaky performance of its buses. As another group of engineers would observe, "even if motor fuel were free, it would still be cheaper to power a trolley or even a streetcar."

So inefficient is the diesel bus that it drives transit users away in droves and transit companies broke. From 1955 (when GM felt its work was done and ceased its overt dieselization) to 1973 over 500 bus companies have abandoned operations, leaving many cities without any transportation alternative to the car. Examined any which way, the diesel bus has no place as a mainstay of an urban center. While the electric bus lasts twenty-five years, the diesel lasts only eighteen. Added to the disadvantage of the shorter life of the diesel is the disadvantage of operating costs that are 40 percent higher.

The difficulty of getting in and out of traffic quickly has given the modern diesel (which is notoriously poor at acceleration) an average rushhour speed of close to twelve miles per hour, or less than the streetcar achieved before the turn of the century. Passengers are also discouraged by the diesel's noise (thirty decibels or 1,000 times louder than the electric's) and foul smoke (more complaints are registered each year over diesel exhaust than any other pollutant).

Yet under the watchful eye of the auto lobby, the lure of the diesel was irresistible. GM didn't have to buy out the transit systems in most Canadian cities as an emotional, anti-streetcar lobby took over. So detested became the streetcar in cities like Montreal that the Montreal Transit Commission refused to either sell or donate its discarded streetcars to the many individuals and organizations who requested them, preferring instead to pay to store them before periodically dismantling and burning them in groups.

By 1955, when Vancouver abandoned its Hasting East line, the last city streetcar route west of Toronto and north of San Francisco went with it.

Toronto's streetcars did not escape unscathed, with a history of intrigues against them that included contrived power failures as a guise for "temporarily" replacing streetcars with diesels along certain routes. Comic relief was provided by the Department of Highways, too obviously eager to share the spoils by snatching the track allowances of streetcars for pavement widening.

TTC commissioners who defended the streetcar were replaced with ones who didn't. When Gordon Secord was appointed TTC Commissioner in the early Sixties, he was praised as a man with "allied transit interests." The allied interests were in automobiles—Secord was the owner of a car rental company, and not unexpectedly, he acted like one by announcing plans to speed up the destruction of the streetcar on the rationale that streetcars cause traffic jams.

Auto-oriented literature was soon produced by the Toronto Transit Commission. One piece proudly proclaimed: "Overnight, streetcars will disappear from Dupont Street, Davenport Road, and all but a short stretch of Bay Street, thus increasing the capacities of these streets at no expense to motorists."

Unlike other North American cities, however, Toronto halted the destruction of the streetcar. To counter the auto lobby, citizen groups formed and soon succeeded in mounting enough public pressure to save the streetcar. As a result of this citizen action, the substitution of diesel buses for electric vehicles was not only stopped but reversed. The trolley bus was reintroduced in downtown Toronto and the diesel displaced. Only two diesel routes remain in the downtown area, and soon there will be one—plans are underway to rerail the once-famous Spadina line, which hasn't seen a streetcar in more than thirty years.

Those who want to imagine how their cities in the future might function can have a preview in Toronto today. During the day, close to 50 percent of all trips city-wide are public transit trips. For rush-hour trips in the core, the percentage approaches 75 percent. Torontonians take an average of one public transit trip every two days, more than the residents of any other North American city.

The trend to electrification is now continent-wide. It is so advantageous that even oil-rich areas like Calgary and Edmonton are giving up the diesel. Edmonton's streetcar system (North America's northernmost) discontinued service in 1951. In 1974, its revival was undertaken, and Edmonton now has added both trolley buses and streetcars, challenging Toronto as the North American leader in public transit.

Virtually every Canadian transit system stayed healthy throughout the Fifties and Sixties. At the start of the Seventies, they were holding their own; by the close of the decade they were all placed on the critical list. Electrification will not be enough to make any North American system healthy, because dieselization has become public transit enemy number two.

General Motors was able to slow down its dieselization program in the mid-Fifties because it saw a new factor emerge that would result in certain, if slow, death for public transit. The same slow disease that has so seriously debilitated American systems has spread to Canada, with the result that Canadian systems are also in danger of dying. The disease has its origins outside the cities it has attacked, and it has taken the ancient Latin name *suburbia.*

The same transportation subsidies that were going from the non-motorist to the motorist were going from the city to the suburb. The city dweller did not need the six-lane suburban highways to get from work to home, but he had to pay for them all the same. The suburbanite would have stood for nothing else when confronted with costs of about one dollar per person per mile.

The city dweller did not need the municipal transit system extended miles beyond the city limit, but with metropolitan governments taking over from city governments, transit systems became further and further extended while fares were flattened so as not to discriminate against the suburbanite.

Municipal transit systems are now the only transportation systems in the world that charge the same fare for a ride anywhere on their system—trains, planes, boats, buses show their bias by increasing fares with distance traveled. The longer the transit trip, the larger the loss. Every new mass transit line from city to suburb becomes a guaranteed loser before it hits the drawing boards.

Suburban routes that stay in the suburbs are also money losers. Not enough people live along bus routes to make transit pay in any North American suburb. For its finances to be different, two to three times as many people would have to live in the suburb: suburban densities would have to approach those of the city.

In 1960, several years after Toronto joined its suburbs to form a metropolitan city government, every one of Toronto's transit routes still made money; twenty-two suburban routes lost money. Were Toronto to leave the Metropolitan government (as has been threatened) it would be the only city in North America to have a self-sufficient system.

Public transit still makes money in the City of Toronto and still props the suburbs up. Every suburban transit commuter receives a subsidy of $110 a year from transit riders who reside in the city. The TTC's best customers—those who keep it in business—are overcharged to support suburban service. The more suburban service is extended, the more it threatens to bring down the entire system.

Transportation priorities need to return to first principles: public transit must be made to pay its own way, which means suburban service must be made to pay it own way. As difficult as this might have seemed to accomplish before the energy crisis, self-sufficient mass transit systems could be a natural consequence of an energy-short world.

The change in the equation will come mainly from the car. With every price rise of gasoline, and every gasoline shortage since the OPEC oil crisis, the use of mass transit has increased. We are in store for price rises, and gasoline shortages, on a far grander scale than those previously experienced. Self-serve gas at three dollars and four dollars a gallon, though cheaper than gas pumped by attendants, may be spared for more essential uses than getting downtown.

Removal of the oil subsidies will be impossible to avoid. As the price of petroleum goes up, so too will the desire to conserve it. Diesel buses will be eliminated from city cores altogether and phased out of suburban use as steam-driven and electric battery buses—also suited to serving low densities—experience a revival.

In a showdown between public and private transportation, each made to pay its own way, public transit would not come out the loser. In Western European and Japanese cities, where the showdown takes place daily, the great majority of urban trips and commuter trips—despite being brutally priced by distance—are public transit trips.

But suburban travel in North America—even by the best of buses—has to remain pricy, and inconvenient, despite computer scheduling and electronic innovations to eliminate long waits at stops. The fault lies not so much with the transit system but the suburbs they're trying to serve.

Like a freak of nature, born misshapen, lacking basic neurological functions, without the necessary life supports but kept alive by intravenous feeding, the suburb had no right to enter this world. But now that it's been created, we can't very well disown it.

# 12
## *The End of Suburbia*

A VISITOR FROM another planet would be most impressed: "The Earthling has a highly developed and highly organized society," he would report back. "His environment has been precisely engineered for his convenience. His transportation system makes accessible all regions of his land area through a network of roads that is well maintained and dedicated solely to his welfare. His social services are placed in establishments beside his roads for his convenience. The establishments provide areas of refreshment and recuperation where Earthlings can be attended to lavishly.

"There appears to be only one deficiency in the Earthling's system: the presence in each Earthling of one or more parasitic organisms called 'man' which are believed to have no function other than to precipitate accidents and keep the population under control."

This report could only have been made in this century. At no other time in human history could the visitor have found that human needs were subservient to the needs of machines, that human activities like walking were limited when in conflict with the car's, that human perceptions, human living patterns, the human use of time and space would have to adapt to relate to this world through the filter of a sixty-mile-per-hour wrap-around windshield.

At no other time in human history have two of man's most basic nonrenewable resources—land and energy—been fused into one package, to be treated as both renewable and disposable. The scale of the squandering of these resources was unprecedented; the natural efficiencies in our ecological system were subverted by the belief that energy's inherent limitations no longer applied. Land and energy were both cheapened. The product of their union is the modern suburb.

Life beyond the city limits is not new. A feature common to all who love cities has been the desire to flee them, to find a place free of congestion and chaos. Not unlike modern man, the weary native of the ancient city of Greater Ur, after a hard week at the office, looked forward to a weekend out of town. In the 4,000 years since then weary urban man has longed for a private place in the country.

Whether castle or cabin, no matter how mortgaged, the home in the suburb was an expression of independence and relaxation. The suburban sentiment that "there is a vast deal of satisfaction in a convenient retreat near the town, where a man is at liberty to do just what he pleases" was not first voiced by an armchair philosopher from Scarborough or Scarsdale but by a surburbanite in the fifteenth century (who also shared with his modern counterpart the disarming luxury of being able "to appear at my door without being completely dressed").

What changed was not man's sentiments but his use of harnessed energy: the transition from horse-drawn carriage to the horseless carriage would put man through his paces.

After 1850 the suburb was not required to remain close to the city. Following the railway track, communities wound their way through the countryside forming around stations three to five miles apart. Shrewd real-estate developers of the day noted this convenient arrangement and promised houses "within easy walking distance of the railroad station."

Around the turn of the century the promoters of the suburbs became the electric-power and transit magnates themselves, the van Sweringens in Cleveland, Insull in Chicago, Beck in Toronto, and Pick in London, England. Contrary to popular belief, it was not the automobile but the Pacific Electric that developed the Greater Los Angeles area, its lines radiating out seventy-five miles—north to San Fernando, east to San Bernardino, and south to Santa Ana. Fifty-six separately incorporated cities, enjoying lush palms, fragrant orange groves and the ocean air had easy access to the concentration of life in L.A.

Suburban development there and elsewhere was extraordinarily efficient, with streetcar corridors lined with shops. Every stop had a complete spectrum of consumer services—grocery stores, banks, libraries, fruit and vegetable markets, churches, drug stores, movie houses. Residential streets that led to the electrified corridor were compactly built (two-storey houses on narrow lots) and close by, discouraging no one from carrying parcels home. On either side of these suburban corridors lay the suburban dream—open fields and farmland, a safe place for children to play, green relief from urban grey. Natural limits had been set by the electric railroad, resulting in tightly knit, closely linked, small-scale neighborhoods, generally with populations under 5,000.

It would not last long. With the close of World War II the assault on the cities and suburbs—postponed by the Depression and war—began in earnest. The cities, particularly in the U.S., would be injured grievously and left for dead. The suburbs would be taken over and their occupants indoctrinated with new customs. The commuting male would learn to value the freedom of being confined to his car on expressways whose rush hour flow regulated his speed to nineteenth-century levels; suburban females with cars acquired the status of chauffeurs while those without were driven to arranging their days around events no more significant than their husbands' comings and goings. Children would be brought up as little more than automatons, programmed to be dependent on adults for transportation and on television for entertainment.

The automobile, in league with cheap land, was the agent of these profound changes. No longer would houses have to be built in orderly fashion, lengthening the streetcar corridor or widening the residential belt. No longer would shops and services need to be provided within walking distance of the home. Now entire subdivisions consisting of nothing but houses could be built on any parcel of land that could be purchased. Now that the homeowner was expected to provide his own transportation, houses could be assembled wherever the politicians could be persuaded to put their expressways.

In the U.S., the suburban phenomenon is best epitomized by the Levittowns that sprang up in Long Island, Pennsylvania, and New Jersey. Built by the Levitt family, the fastest builders in the East, Levittowns were entire communities of thousands of acres, conceived and completed by developers with a reputation for providing "the best house for the money."

Breaking through the humdrum of normal architectural thinking, the Levitts unleashed their genius by introducing the mass-produced suburban house: Levitt simply fitted precut materials together, in assembly-line fashion, on top of a concrete slab on a piece of Levittown land. People who never before had been able to buy a home of their own became the proud possessors of prestige accommodations. By simply eliminating the frills of construction, Levitt could build eighteen homes before lunch and eighteen after, or thirty-six houses per day, giving the Levittowns a distinctive quality others would soon emulate.

After completing his first Levittown in Long Island in 1950, Levitt could look with satisfaction in any direction and see the same square bungalow lined up on both sides of the street, each one sitting on its own 60-by-100-foot lot, each with one beech sapling and one maple sapling, each having a picket fence with exactly eleven pickets. Seventeen thousand homes to his Levittown, complete with every amenity Levitt could imagine any resident should want—schools, parks, shops, swimming pools and recreational centers. Libraries, art galleries, restaurants, movie theaters, cafés, and hospitals were dispensed with.

Spoiling his sense of accomplishment, though, was the intrusion of a huge shopping center on the edge of Levittown, drawing business away from the local shops. His next Levittown came complete with fifty-acre shopping centers at each end of town. More efficient "Master Blocks" of about 900 homes were also invented to streamline the sometimes haphazard assemblage of houses.

Then, what happens to any successful enterprise happened to the Levittowns—certain people, entirely unprovoked, started to be very critical. Levittowns were sterile, they remarked. They were mundane, lifeless, unimaginative, and monotonous.

Although the Levitts did not understand the complaints over their consistency of design, fear of losing their ranking as prestige builders led to reforms designed to silence their critics. Announcing a new Levittown, Levitt said "We are ending the old bugaboo about uniformity . . . In the new Levittown we build all the different houses . . . right next to each other within the same section." From that time on, two- and three-bedroom houses could share the same block.

In future they would push the bounds of diversity even further.

By alternating color and trims, the Levitts found, no more than every one-hundred-and-fiftieth house had to be identical.

To assure the prospective buyer that all kinds of people live in Levittowns, they advertised: "Wanted—Zoologist for a neighbor." To assure him that his neighbors, even if zoologists, would be as boring as he, they launched their "Hi and Middle Fi" promotion, suggesting that while the Levitt house could be rigged up for hi-fi enthusiasts, it was a pretension to be mocked: "You'll find all kinds of music lovers in Levittown," the ad concluded, "but we've discovered that most of them are middle-fi, like ourselves." The advertising worked, attracting the very people it was aimed at, and Levittown became a model for suburbs across the United States of America.

While Canadians were as original as Americans in reproducing the mass produced home, living in a place named "Levittown" seemed too crass for the cultured Canadian taste. Abandoning American artificiality, Canadian suburbs are named to create a mood representing their qualities. There are three moods in Canada. Suburbs can be named either to seem part of a town or a village (Camelot Village) or to seem set well into the country, surrounded by hills, ravines, rivers and, quite possibly, rabbits, foxes, owls, and beavers (Willowdale, Fox Run Estates) or in old English style, to present a feeling of privacy (Old Towne, Meadows of Manvers).

But although there are three types of suburban names, there is only one suburban type.

Don Mills was Canada's first modern suburb. Built by industrialist E. P. Taylor after the war, it has been the shining inspiration for virtually every suburb built in the last thirty years. So revered is Don Mills that no other planning models even exist in Canada. Understand Don Mills and you've understood that what separates the suburbs of Vancouver from the suburbs of Halifax is 3,000 miles.

Take the minimum number of stores and services you can muster and place them in an area called the "town center." Surround the town center with parking spaces and run a ring-road system around them. Surround the ring-road system with high-density housing—

usually apartment buildings—and place blocks of row housing around this area, effectively buffering the single-family housing from the town center. Place winding roads throughout the single-family housing as opposed to the grid network of roads found in cities; surround the entire residential area with arterial roads (making sure to avoid fronting property onto the artery). Now you've created a suburban community that differs hardly at all from any other in the country.

To give the community members a sense of belonging, they need to live in neighborhoods. A neighborhood is formed by taking 6,000 people, putting an asphalt border around them and dropping a public school, flanked by a neighborhood park, in its center. To make sure community members don't feel too neighborly, different classes of people are kept strictly segregated. Single-family dwellings are not designed to allow either roomers or other tenants. Tenants belong in high-rises. The semi-affluent live in the semi-detached homes; the affluent in the detached. To enable people to be independent of their neighbors, every lot is equipped with parking space for a car. The car has made informal meetings between neighbors difficult. Because people drive instead of walk, they try not to bump into neighbors frequently. Informal meetings do take place—there is generally one restaurant and one supermarket—but a hotel or tavern where people can chat over a drink is rare, especially since they'll have to drive to get home.

What originally sold the idea of the suburbs was the absence of congestion. This has come to be equated with the presence of wide roads and green open space. Suburban developers take the green open spaces they're given and put houses in the center of their lots, surrounding them with space but ensuring that the usefulness of the space is destroyed. Only backyards are used by the suburbanite. Front yards function to put distance between him and the fumes of other suburbanites' cars.

The car, known to suburbanites as a menace to small children, is actively discouraged close to suburban homes. Winding roads exist to slow traffic; dead ends and crescents to send it elsewhere. Residential roads represent suburbia's only protest to the car. But, whether praised or protested, the road that the car has created completely dictates the economy and culture of suburbia, completely shapes its architecture and geography.

Suburbia will remain less than "urbia" as long as its structure suits the auto instead of the pedestrian. The street in the city is no mere route for vehicles but an organizing force for the community; pedestrians, passenger vehicles, stores, restaurants, bars, hotels, churches, small and large offices, and libraries are all attracted by its magnetism. The major suburban street, in sorry contrast, has been stripped of all functions except moving cars, making it in form and fact the slow-speed equivalent of an expressway. All it takes to rob a street of its humanity is to widen it to inhuman scale, restricting access to pedestrians and dividing neighborhoods. Slowly the street starts to cater to the car: store-front streets first become roads servicing a series of plazas, and finally arterial roads lined with wire fences, whose human activity turns its back to the activity on the street.

A suburb can now be created without regard for its distance from the city, without needing the intervening land developed. Because of the car, undeveloped parcels of land can be leapfrogged, with high-speed roads providing the link. These new roads, in sordid service to the car, become known as strips. To be seen at forty and fifty miles per hour, signs on the strip must be distortedly large and stripped of all detail. Only the garish can be grasped. City signs, meant to be seen standing still or walking, have no place beside suburban strips that have not allowed for the pedestrians, often not even to the extent of providing sidewalks.

The story of the suburb is the story of specialization and centralization. For the first time, work and home have been strictly separated. Neither city nor farm ever sought to so segregate the two. They never learned the logic of suburban economies.

To achieve economies of scale, developers built super blocks, row upon row of uniform housing that could be marketed with one strategy, to one type of person in one income range. To provide the services the suburbanite would need, developers built the super shopping center—large regional centers, malls enclosing slotted-in stores that could be marketed with one strategy, to one type of organization in one financial status. These regional shopping centers are generally sprawling windowless structures on sites ranging from fifty to one hundred acres, located near expressways and resembling airports: remote and warehouse-like, the surrounding grounds have all been paved over for the landing of long-range vehicular traffic.

For every 1,000 square feet of shop area inside there are five-and-one-half parking spaces outside. To separate inside from outside the window has been dispensed with. "It is nonsensical to have a wall of windows just to see outside," one of Canada's largest leasing agents for shopping centers explains. "Everything that might have been outside has been bulldozed and paved; basically you've just got a sea of cars that you're trying to hide. So you put in your amenities, you put in your courts, you manufacture an outside environment in the courts outside the department stores."

In the city, shoppers like to walk from one department store to the other browsing at the shops in between. In the suburbs, the developer likes to put a department store at each end of his shopping center with shops in between. But unlike the inexpensive city street, which does not need to be financed, built, and maintained at high cost, the economics of the shopping center virtually close it to all but chain stores.

Mortgage financing, developers have found out, comes much more easily with national chain stores lined up as tenants. To get the chains, developers often pre-sign their leases, usually giving rent reductions in advance of obtaining the mortgage money. The independent has the opportunity to rent only those locations the chains don't want, and at unfavorable rents.

By taking all commerce away from the suburban street, the developer has left it depressingly drab, doomed to stay devoid of any pedestrian purpose. By putting all commerce into the suburban mall, the developer has sterilized the spontaneity of the city street, exchanging the hum of the marketplace for the numbing noise of Muzak; he has replaced the jumble of buildings built at different times in different styles for the orderliness that comes of mass production. There are no beggars in shopping centers, no street musicians, no litter, no threat of violence, no signs of decay, no signs of rebirth, no breezes, no sunlight, no chance of rain—no intrusions of life. Suburban life has been made subhuman and suburbanites are the losers because of it.

By separating work from play and rejecting the real for the unreal, the suburb dreamt of becoming the playground escape from the urban nightmare. Concerns shifted from the social to the self. A bigger house, a better car, the next vacation became prime preoccupations. Leisure increased as play became the serious business of life. Replacing the variety of the city were golf courses and country clubs, swimming pools and cocktail parties. In retaliation against the concentration of the city, the suburb became an overspecialized community, more and more committed to relaxation and play as ends in themselves. Compulsive work led to compulsive play, with no gain either in freedom or variety.

No longer having to face life in the city, which becomes little more than a weekday destination for migrant workers, people shun responsibility there. Not wanting to trade responsibility in the city for responsibility in the suburb, they take on few obligations in the home environment either. Social involvement is more or less confined to social agencies, political involvement to voting day. Citizenship is renounced in both city and suburb.

In urban environments, when real or imagined injustices occurred, it never took long for people to band together to try to find solutions, agitate and form lobbies, hold meetings and draw up petitions, demonstrate with placards and form organizations. Housing legislation, libraries, soup kitchens, and hospitals were created through efforts that involved and benefitted the whole community. In suburban environments, people live and die in deluded support of an innocent world, their idyll marred only by unpleasant headlines passed enroute to the entertainment section of the newspaper. The range of human emotion, the passion of the urbanite in defending a city park or a rundown tenement from the demolition crew's hammers, is inexplicable to the suburbanite who cannot imagine himself defending the local shopping center or laying down his life over the cancellation of an expressway project.

The suburbanite is cut off from all but his escape fantasy, alienated from the rest of society, frustrated because of it, having many of his human attributes suppressed, and, even if only vaguely, uncomfortably aware of it.

His loss has now been passed on to his children. Formal education, always a supplement to the rich and varied experiences available to a child in a small town or big city—where industry and the workings of life are not camouflaged—becomes inadequate. Suburban education faces the impossibility of trying to put information in perspective for a child whose world view has seen all non-suburban phenomena interpreted through the TV screen. The suburban child grows up armed only for subsistence, suburban survival—an indirect consequence of the direct interference of governments.

That the city was deserted for the suburb at all was not so much because the car made it possible but because the state made it mandatory. The suburb was a government offer the North American could not refuse.

Faced with a housing shortage in 1945 with the postwar return of servicemen, governments in Canada and the U.S. passed a flurry of legislation—the National Housing Act, the Federal Housing Act, and the Veterans Act—to insure loans at up to 95 percent of the value of the house, tax write-offs to builders, incentives to build expressways, and amendments to the Insurance Act to allow insurance companies to finance land improvements in suburban areas.

Without the FHA in the U.S. and the NHA in Canada, neither Levittown nor Don Mills would have been built; the banks would not have taken the uninsured mortgages of young middle-income home buyers whose long-term financial status was not secure. But with this legislation—guaranteeing a long-term housing market—it became easy for both buyer and builder to obtain loans. Its effect was to drive the price of owning a home below the price of renting one, and it set off the greatest mass migration in North American history, topping the trek to the west of the mid-nineteenth century and the great flood of immigrants from overseas a half-century later.

By the mid-Sixties the metropolitan areas of the continent had become predominantly suburban and the suburbanite had become the most heavily subsidized individual on earth; the envy of every welfare bum. The subsidies were so great that they obscured information we are only now discovering: that life in the suburbs has never paid its way. What makes suburbs uneconomical is precisely what has also made them attractive—their sparse population.

Low-density living does more than make public transportation a losing proposition. It makes virtually every other public service costly as well. In some Canadian municipalities each new suburban house will produce a tax-dollar deficit as high as $900, forcing mill rates to be increased for everyone.

Logically, low densities should be expensive. Fire, police, and ambulance service, for example, must be available within three or four minutes of an emergency call, regardless of the number of people served. When one-half to one-third as many people live in an area, the cost of providing these services will be two and three times as great. Low densities literally cause the burial of vast amounts of unnecessary capital. A six-inch water main is required for fire-fighting purposes, even though a typical suburban street might only need a two-inch service for all its residential needs.

The additional costs alone amount to $6,000 to $8,000 per lot over what could be attained by putting the same houses that are on standard suburban lots on smaller but more useful lots that would allow for increased landscaping and private screening, more park space landscaped for children's play, and improved control over the design and siting of individual units. The conventional suburban density would double to about twelve-and-one-half houses to the acre. This would result in about fifty people to the acre, compared to the standard of eighty people per acre in the older developed areas.

As one study shows, even tripling the density would not adversely affect claustrophobes. No building, including apartment buildings, need exceed three stories in height and three times as much park space could be provided. Forty percent less road space would be used.

Realization of the mounting costs of low-density suburbs is encouraging suburban governments to slough off costs, charging fees for participation in publicly run programs that were previously free. In the Toronto suburb of Scarborough, local residents have become responsible for equipping local parks with whatever swings, seesaws, and slides they feel their children deserve.

One suburb, which was particularly poorly planned, has actually experienced a fall in prices. Since October 1, 1947, when its first house was occupied, the original selling price of $6,990 has risen steadily to the current average of $50,000. But in 1977, the average selling price slipped $26 from the previous year while house prices elsewhere were still climbing because of inflation. That was the first time that house values had dropped in Levittown's thirty-year-history.

The twentieth century suburb was first conceived around the turn of the century, not as our senseless sprawl—an indistinct suburban belt—but as satellite cities of 25,000. "New towns" around major cities, they were to be surrounded by green belts and linked to their cities and themselves by the complex transportation system then available. But instead of maintaining our transportation choices, which matched alternative routes, vehicles, and speeds to individual preferences, we have become singularly dependent on the automobile, which, when given the license to expand willy-nilly, expropriated the one commodity the suburb truly offered: space. The result is a perversion of the suburban ideal. Instead of buildings set in a park, we now have houses set in a parking lot. In the rush to claim the private plots, the privacy became a mockery, no longer worth acquiring, a dream paid for out of the pockets of the non-dreamers.

The damage already done is largely irreversible and incalculable. Much of the best farmland on the continent has been paved over so that suburbs could grow. An experience-impoverished generation is now entering adulthood, ill-equipped in the social resources to assume the responsibilities needed in a world short of physical resources. As energy prices continue to rise the suburb will be increasingly seen as a burden not worth bearing. Suburbanites will be pressured to pay more of their own costs. How this will be done is critical.

To remove all subsidies, to make suburbia pay its own way right away would create severe personal hardship and social chaos. To leave the structure of suburbia alone would only continue the drain on non-suburbanites while perpetuating the problems of the suburbs.

Subsidies to new developments could be withdrawn entirely. This would minimize additional sprawl and phase out transportation costs, beginning a profound shift in living patterns in and around suburban communities. Higher energy costs will persuade people to work closer to home when next considering a move, to seek innovative transit alternatives like jitney services and car pools, to zone for higher densities, particularly near rapid transit facilities. The boundaries of individual suburbs will become distinct as they contract and re-form around transit corridors—recreating the patterns of density prevalent before the automobile dominated travel.

Immediately beyond the high-density transit corridors, however, field and farm will no longer exist, only today's stocks of (for the most part) substandardly built suburban homes. The very posh, well-built areas may assume the status of estates and prosper. But the homes in most areas will prove too costly to maintain, especially those homes deepest in their subdivisions. Inaccessible to public transit, the cul-de-sacs and crescents so highly prized today will be the first to be deserted for urban pastures, the first to fall into disrepair, the first to form local slums and fall before the bulldozers. Many suburbs will fare well in regrouped form, many will disappear altogether, being turned to other uses. Some land will be suited for recreational purposes, providing cities with less remote locations for the golf courses, zoos, and amusement parks that are presently expropriating our diminishing agricultural lands and wilderness areas. Some suburbs will reform around the local resources so necessary to any community's *raison d'être,* or become industrial sites for the new products and services the city will need. And some will be returned to agricultural purposes, feeding themselves and their cities as they fed previous generations living on the land.

Beyond today's suburbia the resource regions upon which our cities now depend will become doubly resourceful, cashing in on their energy potential. The livestock, crops, and logging operations of the future will yield, as a byproduct, vast quantities of methane gas (from animal wastes), ethanol and methanol (from crop and forest wastes) that can be piped to cities, often through existing networks.

In 1978 Chicago's People's Gas Company began to buy manure-derived methane from several large cattle feedlots near Guymon, Oklahoma. Each day, up to 1 million cubic feet of methane is pumped into the same pipeline carrying natural gas from Oklahoma to Chicago. For added income, the feedlots sell the residue from the extraction of methane gas as fertilizer and cattle feed.

Country and city both benefit from the increase in efficiency.

In the process of restructuring the suburb, the supremacy of the city as the center of culture, commerce, and residential life will also be reclaimed. For the city is no mere industrial machine but an organic entity that's wholly dependent on diversity for its dynamic equilibrium. When the city has previously come under attack—such as during the deplorable conditions of the Industrial Revolution—coal was the urban agent of change responsible for stripping the city of much of its diversity and adaptability. Today the city is under a potentially much greater attack due to another energy form—nuclear power—a far greater centralizing force than coal. While coal can be stored or used in small amounts, and can be transported great distances, nuclear power is strictly an urban phenomenon. It makes such great and constant amounts of power available to its vicinity that only the large concentration of people found in cities can efficiently make use of it.

Cities have been centralized before, but their diversity has always re-emerged. That they have survived 6,000 years of wars, migrations, disease, and degradation is testimony not only to their strength as an institution but to their claim to be our most efficient forum of human communication.

# 13

## *Necropolis*

THERE IS ONLY one way to move great numbers of people with great speed and efficiency in cities: by foot. One hundred thousand pedestrians could manage the journey from downtown Boston to the Common, using the existing streets, in a half hour with ease. Give them the advantage of the automobile and only hundreds would complete the journey. Before there was any public transportation in England, about 50,000 people per hour used to pass over London Bridge on their way to work. Our best expressways today manage 4,000 to 6,000 cars.

While urban expressways permit drivers to make longer trips, the car, even in an auto-oriented city like Los Angeles, can't compete for speed with the pedestrian in the more dense transit-oriented cities. Pedestrians can reach more destinations within a five-minute walk or an elevator ride than motorists in twice that time. By their nature cities are suited for people who, it is important to remember, have needs different from machines. Machines do not need a good night's sleep to function well, do not increase their productivity when they're patted on their posteriors, do not value the time they're off work for minor repairs.

Unlike machines, humans like to start work in the mornings, eat all their lunches between twelve and two, and take most of their vacations in the summer. If their machines must lie idle most of the time to suit their whims, that's the way they'll have it. Transit systems will be patient between rush hour demands; restaurants will wait until the tummies of their customers growl.

As long as cities are designed for humans—inefficient though humans may be—the city will be remarkably efficient. Make the city over to accommodate the machine and the efficiency is lost and the city threatened for both man and machine.

Machines have always brought cities to the brink, making yo-yos out of human inhabitants, reeling them in and spinning them out with each new form of energy tapped. When water mills became the vogue in the sixteenth century they attracted industries out of towns to regions where small, swift-running streams or waterfalls could power textile mills and other enterprises. Human populations followed the industries to the countryside. By the eighteenth century this power source (aided by cheap land and cheap rural labor) had shifted commercial power to the village factory, which would soon educate the city to the economic advantages of the factory system. It was in the verdant valleys that the misery of monotonous work, child labor, and the twenty-hour-day would be institutionalized; and underground mining, formerly considered punishment for criminals, would gainfully employ the unconvicted.

But although village industries could be vicious to villagers, they had the virtue of being small scale. Much as they might have liked to devastate the natural environment as enthusiastically as they were devastating the human environment, local mining and smelting, water power, and the canal transportation system of the day did little to rearrange the landscape. Until the nineteenth century, industry remained decentralized in small workshops, scaled to agriculture.

James Watt and his steam engine changed all this: the power of the piston packed industries around the coal mines that fueled it and changed the scale and nature of the city. It now became possible to destroy both man and his environment.

Steam worked best in big concentrated units. For power to be delivered efficiently through the belts and shafts worked by the central steam engine, every spinning machine or loom had to be located no more than a quarter-mile from the power center. The more units packed within that quarter-mile radius, the more profit the plant could yield. The small-scale village factory gave way to factories employing 250 laborers. A dozen such factories, with all the services needed to sustain their still serflike staffs, was now a good-sized town.

Profits could be increased not only by more efficient use of steam power but also by less humane use of human power. Keeping wages low was most easily done in the cities with their large and pathetically unexploited labor pools (especially after the towns' supply of child labor dried up due to government meddling in the marketplace). Man now lived in congested cities around the coal fields, in congested cities that could be conveniently fueled with their coal, and in congested pockets along the train lines linking the two. The misery was now spread more equitably.

Lifestyles changed. Reviving a custom that had been outlawed in many English towns in the sixteenth century, it became acceptable once more to dump garbage into the streets where it would stay, no matter how vile and disgusting, until the accumulation enticed the entrepreneur to carry it away for manure. Manure, however, was never in short supply. A study in 1845 of 7,000 residents of Manchester found less than one toilet for every 212 people—foul beyond description and usually kept in the cellar next to the pigs. The dirt and congestion brought rats, bedbugs, and lice that transmitted disease and discomfort.

In the latter half of the last century more than man's creature comforts changed. Man became, in fact, a different animal. Working ceaselessly in subterranean, subhuman conditions, he learned to think of malnutrition as victory over starvation, of lost limbs and chronic disease as victory over death. Almost understanding that a race of defectives was being created, he became resigned to malformations of the bone structure and organs; rickets in children due to the absence of sunlight; defective functioning of the endocrines due to the diet; skin diseases for lack of water to wash with; smallpox, typhoid, scarlet fever, and septic sore throat from dirt and excrement; tuberculosis, bronchitis and pneumonia because of bad diet, sunlessness, and overcrowding—all nurtured in an atmosphere stenched with chlorine, ammonia, carbon monoxide, phosphoric acid, fluorine, methane and some 200 cancer-causing chemicals. Being a party to this, either as victim or victimizer, living always in soot and grime and filth, having all norms of his physical existence changed, changed all norms of his metaphysical existence. Debased, man's taste for fine things degenerated.

Black smoke and soot made black clothing practical and the black stovepipe hat a fitting symbol: the acids and chemicals in the air affected the throat and lungs and lowered the general tone of speech and taste; the thick musty odors of coal and chemicals blunted appreciation of subtler smells.

The internal environment became as impoverished as the external one: grotesque wall paper disgraced walls and tasteless furniture littered floors. Religious relics reached an esthetic level approaching profanity.

Diets changed. Even the well-to-do ate canned goods and stale foods when fresh ones were available, no longer being able to distinguish between the two. Art changed. The bright colors preferred by painters of a previous age were thought unnatural and unartistic by critics who knew nothing other than the greyed monotony of their surroundings.

Never before had man enjoyed such a high level of energy consumption.

Man was made over to the machine; the *status quo* became the always was; technological competence was used to obscure social incompetence. The deafening pound of the steam engine was not only tolerated but encouraged as a symbol of power and production: Watt was forbidden by the manufacturers to reduce the noise of his machines for marketing reasons.

Apologists for industry excused the sacrifice of society to the god of progress: high mortality rates for those sharing in industrialization were discounted or ignored; critics of the health dangers of pollutants were branded as alarmists; standard-of-living statistics were cited as improvements in the quality of life and the importance of all environmental factors was made subservient to the importance of the economy. Even the sun—indispensable to human processes—was seen as a commodity to be replaced when a better technological solution presented itself.

The evidence of physicians, testifying that the wide unavailability of sunlight in the smoke-bound factory communities was retarding the growth of children, was pooh-poohed with claims that the invention of artificial illuminating gas—used as lighting in factories—made the sun irrelevant.

It was a society led by (instead of leading) its technology.

The most degraded urban environment the world had yet produced, where even the quarters of the ruling class were befouled, was not recognized for what it was by a populace conditioned to accept the authority and judgment of others. Progress would be slow, with the city solving many of its problems only toward the end of the nineteenth century. With the coming of the car, problems on a different scale—but based on the same infatuation with the needs of machines—would present themselves. The car was going to decentralize society—the same way electricity plants were going to relieve congestion by transmitting cheap energy to plants relocated outside downtown cores.

The result was only the timed transfer of congestion from suburb to city each morning and evening and the extension of the city's poisons to the country. Much of the acid rain that is murdering our lakes and killing our crops—in effect irreversibly undermining our entire existence—comes from the car, the balance from the industries that feed on our cities and the electricity plants that fuel them.

Matching in scale the increase in damage we are doing ourselves is the increase in scale of our self-deceit: the potentially catastrophic danger of continuing to heat up our environment by burning fossil fuels; the cancers we will be unleashing on ourselves from the toxic chemicals buried but making their way to us through the water system; the cancers to come from the continuous releases of radioactivity from uranium mining and the normal operations of nuclear power plants are all matched by correspondingly confident assurances from our government leaders and industry-hired scientific experts whom we find all too comforting to question.

The prospect of these slow deaths not being enough, new forms of annihilation—nuclear weaponry—have been devised. There is a nuclear bomb targeted for every major North American city.

It is impossible to predict the likelihood of a thermonuclear war but, whatever the odds, few dispute that it must be avoided at virtually any cost. Every step in that direction, every reduction in the nuclear arsenals of the world—no matter how small or symbolic—is greeted with a sigh of relief by politicians and public alike.

Yet at the same time that we are desperately working to remove the nuclear missiles being trained on our cities by foreign powers, we ourselves are training nuclear power on our own great cities—not deliberately, but as an inescapable outcome of our policies. New York City, Washington, Miami, Sacramento, Cleveland, Montreal, Toronto and others are all within range of the nuclear power plants we've built. Each reactor contains the radioactivity of 1,000

Hiroshima bombs, with the potential to wreak incredible havoc should the worst possible nuclear accident—a meltdown—occur.

A meltdown would be the equivalent of a very "dirty" nuclear bomb. Weapons are designed to release their energy in a sudden blast, sacrificing slow radiation-poisoning for great initial destructive power. Maximum heat emission is reached at four-and-a-half seconds after detonation; half the heat is released by the end of ten seconds. A commercial nuclear reactor, however, with no pressing problems of war, does not mind taking years to kill and maim its contaminees. According to an exhaustive study commissioned by the nuclear industry, a major nuclear accident could result in about 45,000 cases of radiation sickness requiring hospitalization (of which only 3,300 would die) followed by 45,000 fatal cancers and 250,000 non-fatal cancers in the thirty-year period following the accident. Each year, the population surviving the catastrophe could expect to produce 170 genetically defective children.

The City of Toronto has four nuclear reactors operating in its vicinity with another eight being constructed within striking distance of its population. According to a royal commission report studying nuclear safety, the chances of a meltdown at one of the twelve nuclear reactors near Toronto during their thirty-year lifespans is about one in fifty.

Should a meltdown occur and the wind be blowing toward Toronto, in addition to the human casualties the city could be made uninhabitable for 500 years, crippling the provincial economy in the process. Even with the co-operation of the winds, the result of a meltdown at any of those plants could be the poisoning of the Great Lakes, making all the cities ringing them uninhabitable for centuries.

No nuclear bomb has come as close to wiping out North American cities as the Fermi I nuclear reactor in Monroe, Michigan, located thirty-five miles away from Detroit and Windsor. In 1966 it suffered a partial core meltdown and came within a razor's edge of costing us a major city. The runaway reactor was brought under control, but only after so much damage had occurred it became permanently radioactive, needing to be entombed in concrete and guarded for the next 240,000 years. Beside this monument to blind human ambition is being built a monument to blatant human arrogance—the Fermi II reactor.

Our preoccupation with nuclear power knows no bounds and defies all reason. The intense efforts by governments to reduce the threat of foreign bombs—over which we have no control—are matched only by the intensity of their efforts to increase the number of domestic reactors over which we have absolute control. Ironically, should governments succeed in accomplishing both aims—elimination of all offensive nuclear arms and the massive buildup of nuclear power plants—we would be no more secure from nuclear war. Our enemies would now need only to land a conventional bomb on one of our nuclear reactors to effectively detonate a nuclear bomb. Our cities can now be besieged by our own time bombs awaiting only an enemy timetable.

The attack need not come from a foreign power. Terrorist groups in North America are not unknown and nuclear power plants—the very symbol of the industrial establishment they despise—become a target of attack, allowing terrorists the luxury of crippling a sector of society without directly punishing those they claim to speak for, as throwing poison into the water supply might.

We already have about one hundred nuclear power reactors in North America, with hundreds more on the drawing boards. If built, there would be thousands of shipments of radioactive fuel per year along hundreds of thousands of miles of route, all needing to be heavily protected against saboteurs. Workers inside plants would want other workers to be screened for mental and political instability. Citizens in all nuclear-powered cities, and along all transportation corridors, would need the assurance that newcomers and passers-by were always where they were supposed to be.

Personal confidentiality would become a petty consideration when the physical survival of society was concerned. A complete suspension of all civil liberties, if not seen as necessary before the first act of sabotage, is a certainty as soon as society appreciates the alternative. There was no sense of injustice at mandatory personal searches in airports during the era of hijackings, despite the relatively few deaths that accompanied the airplane dramas. In a nuclear-powered society, with the stakes so much greater, public outrage will only be expressed when the personal surveillance is not strict enough.

Those who are promoting a nuclear-powered future—the all-electric society—are planning nuclear reactors for virtually every

North American city. All who live in cities or travel through them, all who live along their transportation corridors or travel along them would become necessary suspects.

Human relations, communications, and travel patterns, would all have to conform to suit the requirements of our energy forms and the machines they require. When man can meet the needs of his machines he will be considered efficient, when he can't he will be considered a waste product: unemployable or deviant or dangerous.

The lack of tolerance in the nuclear machine, where undetectably small mistakes can cause incalculably great consequences, will not allow for experimentation or independence or diversity. Under a nuclear-powered system we will end innovation and evolution, become ingrown and restrictive, insecure and protective. And, as in all highly centralized systems, our social system will become increasingly and inevitably vulnerable to collapse.

# 14

# *Journey Into the Center of the City*

THE JOKE'S ON US. The mass migrations into and out of the cities, the 50,000 deaths per year from the automobile, the moral and physical devastation of suburban development, the degrading, diseased urban conditions of the Industrial Revolution, the destruction of our inner cities, the desire to decongest and decentralize cities, the industrial rush out-of-town, the reaction against the city, the back-to-the-land sentiment, the longing for the simpler life, the search for our roots, our soul, our *raison d'être*, all these have been based on the premise that the source of everything is the land, that cities are parasites feeding on their surrounding wealth, that cities were not possible until agricultural development could support them.

On this last point Karl Marx and Adam Smith, favored economists of the left and right, would both agree. Nor would many others—lamenting some bygone era—question the idea that digging toes into rural dirt is preferable to having them soled for the hard surfaces of the city. The notion that we are happier, healthier, and closer to God in the country appears in the writings of Thoreau, Rousseau, Mumford, and in the works of Le Corbusier and Frank Lloyd Wright.

Though few seriously maintain that cities have no important role to play in society (Cambodia's Pol Pot being a notable exception, as witnessed by his forced evacuation of Phnom Penh), much of our present thinking continues to flow from unstated assumptions derived from, and limited by, this theory of urban dependence. As a result, our solutions to urban problems have tended to be one-directional. The direction has been toward making cities greener, more spacious by making them less congested, less chaotic by making them more orderly, less dirty by relocating industries out of sight—in other words, more like the countryside. There have been few championing the other direction: urging cities to strive for higher densities of activity, and less dependence on rural resources.

For the city's potential to be realized, the city must be seen afresh; the intellectual baggage associated with its origin must be discarded. At the very least, the city has always been second nature to man, and if Jane Jacobs is correct in her *The Economy of Cities*, the city actually preceded rural development. Whether chicken or egg, the city's history as its own provider is an impressive one. Cities have not only been industrial and commercial centers but were even the origin of agricultural practices, inventing and reinventing farming technologies.

The corn-fed steak sprang out of the stockyards of Chicago and Kansas City and only later moved out of the city, much as the meatpacking industry is today moving out, when the city has better uses for the space in its confines. New seed stocks, new agricultural implements, new agricultural concepts have always arisen in the urban areas.

The Egyptians of the Old Kingdom force-fed hyenas until fat enough for slaughter, kept pelicans to lay eggs, tamed mongooses to kill rats and mice in their granaries. Their antelopes wore collars, their asses were domesticated, and all within their city limits.

The early medieval cities, great commercial food centers (chiefly wild fowl and fish), actually exported their surplus great distances to feed many other little trading cities and manors in the countryside. Some of the grain these cities used probably came from rural areas but most of it did not. In the fields within and outside its walls the common practice of the city was to grow its own grain, which was then processed in city mills and bakeries only later to be incorporated into the village and manorial mills and bakeries.

What historians call "the twelfth century agricultural revolution" (a scheme of crop rotation that improved on earlier soil-depleting practices) centered around the cities, spreading first to the rural areas nearby and taking more than 200 years to reach rural backwaters. The "eighteenth-century agricultural revolution" (when fodder crops like alfalfa and clover were first introduced into the crop rotation, providing the nitrogen necessary for growing grain) actually started a full century earlier in the gardens and fields of French cities.

Hybrid corn—a revolutionary change in American agriculture—and hybrid wheat—revolutionary for the cold Canadian prairies or for the heavy rains and soils of Quebec—came not from the corn or wheat farmers but from scientists in plant laboratories. The shift in New York State from wheat to fruit farming originated in a nursery in Rochester, New York. The fruit- and vegetable-growing industries in California did not evolve from the state's older wheat fields and animal pastures but came full bloom from the marketing concerns in San Francisco. Fish farming in Manitoba and lobster farming on Canada's West Coast came from the ingenuity of urban thinkers.

The process continues today with new seeds being developed in Toronto, calf vaccines being developed in Saskatoon, and new concepts in agriculture—the energy crops that will be providing hybrid grains for gasohol and hybrid trees for methanol—coming from the urban energy capitals of North America.

Innovations in agricultural materials and equipment also came from the city, originally less for export to the rural farm that for use in cultivating city crops. Though many of these functions have now been transplanted to the country in the form of fertilizer factories, agricultural research stations, and tractor plants, their urban origin is clear. Mechanical sowers, cultivators, harvesters, irrigation equipment, canning, freezing and drying technologies and more are examples of the transplant of urban techniques to rural applications. McCormick's horse-drawn reaper, an innovative farm machine that replaced a complex manual operation, adapted devices already in common use in industrial work.

That rural development follows city development can be seen almost anywhere in the world, at almost any period in history. Countries with the best agriculture performance also have the best industrial development.

Adam Smith noted the contrast between the backward agriculture of agricultural countries like eighteenth century Poland and the progressive farm practices used in industrial England. He went on to report that it is not agriculture that leads the way for the development of industry and commerce, noting the superiority of English industry and commerce over English agriculture, and that the most productive, prosperous, and modern agriculture was to be found near cities.

Parallel observations are available to us today by contrasting Western Europe's industrial and agricultural technologies with those of Eastern Europe, or by noting that Japanese agriculture thrives today alongside Japanese industry while before the war Japan's agricultural development lagged well behind, despite the industrious and thrifty character of the Japanese farmer. Before the war one-quarter of Japan's rice needed to be imported due to its small supply of arable land. After the war, with infusions of urban technology, the rice deficit was eliminated. By 1960, though the population was 25 percent larger and total consumption of rice had soared over pre-World War II figures, Japan was not only self-sufficient in rice but producing far more milk, fowls, eggs, meat, fruit, and vegetables for the urban populations that were providing the means for their production.

The notion that the city is subservient to its surroundings is deeply ingrained despite the evidence of the twentieth century, or the observation of Marx in the nineteenth, and Smith in the eighteenth. We continue to act as if cities are dispensable, as if city lifestyles are artificial, as if city functions are uneconomic. In doing so we fail to realize that the city has always been sovereign, remarkably so, with invisible efficiencies of such strength that it has rarely faltered, despite being under near-continual philosophical or physical attack, despite seeming, by all standards of measurement, the paragon of inefficiency.

Corner grocery stores may seem inefficient next to large centralized supermarkets stocking not only food but hardware and magazines as well; the renovation of old tenements may seem less efficient than tearing them down and replacing them with apartment blocks housing several times the number of families; it may seem inefficient to keep industries in cities when the suburbs can provide cheap land and convenient roads for pickups and deliveries.

Yet these apparent inefficiencies are not deficiencies, but the source of unseen efficiencies and the fountainheads of innovation. The grocery stores of Montreal, all but eradicated by the supermarket chains in the Fifties, were institutions that provided far more variety, service, and flexibility than the supermarkets. Because they were conveniently close to residences, fresh food could be picked up on the way home. Freezers were unnecessary and virtually unknown in Montreal's neighborhoods and chemical preservatives did not need to be used on a grand scale. Because regular deliveries could be easily and cheaply made by youths on bicycles, the less-active elderly did not need to be placed in old folks homes, and no house required today's large storage areas for food. Each customer represented a substantial amount of business for each corner store, encouraging individual needs to be met. Phone orders were common. Stocking a certain kind of imported candy or carrying a particular brand of sausage at the request of customers was also easily accomplished, and though the purchase price of specialty items was often higher, they were at least available for those who wanted them.

These additions to the corner store's stock were, of course, encouraged as an important new source of business. Other customers would try the new products, opening new markets for items that would be otherwise unable to establish themselves. Local innovators, not needing to start big, could also sell their wares directly to the corner stores. Home-made jams, pickles, baked goods, products that later led to fair-sized operations, started this way with the result that cheeses and breads produced nowhere else can be found in Montreal. New enterprises sprang from old, providing a vitality and variety for enterprises and the enterprising. Even today, Montreal's corner shops have retained some of their immense diversity: the small grocery stores in one neighborhood stock different products from the corner stores in others. Measured not by the different brands of cereals and soap detergents but by the different varieties of fresh and canned produce, imported goods, distinctive meats, the Montrealer living in the center of the city has access to more goods within a ten-minute walk of his home than the Montreal suburbanite within a half-hour drive.

It is precisely through the process of re-fashioning old work to meet new needs that new work is created. Progress comes not from doing more of the same but from finding better things to do.

When the 3M Company, one of the world's great innovators, began in Minneapolis in 1902 it consisted of a handful of people in the mundane business of digging, crushing, sorting and selling sand to manufacturers of metal castings. Their first great leap forward—

sticking sand to paper to make sandpaper for carpenters and cabinetmakers (not even an original idea)—was frustrated when they couldn't get sand to stick properly. In the process they invented a masking tape for house painters which led to (in order of emergence) shoe tape, electrical tape, acetate tape, pressure-sensitive adhesive tape (better known as Scotch tape), acetate fiber tape, cellophane tape, printed cellophane tape, plastic tape, filament tape, sound recording magnetic tape, and nonwoven synthetic fibers, in addition to dozens of other families of adhesives and uses for sand.

What 3M did was find new uses for old products. They began by supplying sand to metal workers and then diversified by supplying sand to woodworkers. The adhesive they tried to develop for woodworkers led to adhesives for painters and electricians. Today, new uses for sand are still being developed, while old technologies are being preserved.

By adding new work to old the sailboat has survived. In recreational form, there are more sailboats today than in the age of sail—the sailboat has built new technologies upon old. This preserved knowledge of boats will soon reap additional economic benefits as the wind-powered freight-carrying vessels of the future become adapted from the sailing ships of the past.

In similar if less spectacular fashion, hundreds of innovators each day devise new ways to cut hair, clean clothes, or add numbers. New goods, new processes, and new services result, renewing cities in the process. Where this innovation does not occur, cities stagnate. Although cities may often seem to be progressing according to grand designs like world's fairs and Disneylands, they are in fact only acquiring the efficiency and inevitable obsolescence of a machine—a good producer as long as it's current, but scrap material as soon as its product is no longer needed or a more efficient machine comes along. Such models of efficiency—one-industry towns—invariably decay and become ghost towns unless their fall is delayed by outside props like government grants.

Detroit before 1920 seemed extraordinarily inefficient, with hundreds of enterprises trying to produce automobiles and thousands of parts manufacturers duplicating efforts. As the automotive industry became centralized and "efficient" Detroit assumed the same kind of efficiency, making itself over entirely to the needs of the machine it had become. As its industrial efficiency increased, so too did its invisible efficiency decline. Detroit became the best North American example of a socially diseased city, as stable, sensible, and sound as its automotive industry.

Manchester, the city on which Karl Marx based much of his analysis of capitalism, was the most talked of city of the mid-1800s, acknowledged by all as the city of the future. Its huge textile mills, operating at breakneck speed and efficiency, dominated the city. Birmingham, when noticed at all, was viewed as an anachronism, a city that somehow seemed to muddle along despite its inefficiencies: it had few large industries; most of its manufacturing was carried out in small organizations of a dozen employees or less. The organizations weren't consolidated; their work overlapped and they wasted motion. Even these organizations seemed unstable, with their best workmen breaking away to start new businesses that only served to add to the duplication.

Manchester's specialization led to its obsolescence. As people in other places learned to spin and weave cotton efficiently, Manchester lost its markets and, stripped of earlier skills, sank into decay. Its populations left for Birmingham, London, and other cities where there were still opportunities to be had.

Birmingham prospered where Manchester failed because of its ability to turn old technologies into new, because it crossbred enough skills within its walls to give birth to new progeny. To its saddle-and-harness-making industry was added related hardware and tool manufacturing. When shoelaces put an end to its shoe buckle trade, its button industry compensated for the loss. The use of glass in buttons led to a considerable local glass industry. Diversity was the source of Birmingham's resilience; specialization was the source of Manchester's and Detroit's rigidity and vulnerability to changing conditions.

Specialization, by definition, will destroy any system, any city, over time. Mohenjodaro and Harappa, the twin capitals of the ancient empire of the Indus, achieved remarkable levels of production in standardized bricks, turning them out endlessly in multiples and fractions of sixteen. Their mass-production techniques also spewed out so many identical pottery cups that it is thought they developed the custom of drinking, then smashing their cups. But this output in production accompanied stagnation in innovation, and sometime before the year 2,500 B.C. developmental work had halted—no new goods, no new services, no improvements are recorded from that time on, only the mindless mass production that led to their downfall during the same period other people prospered.

In the same way that North American industry, rather than produce new work, relied on planned obsolescence to keep output high, the economy of Mohenjodaro and Harappa opted for unproductive work. Both mistook volume for productivity; both let their economies become highly centralized and oriented to mass production; both thought as would Adam Smith, that the division of labor, in itself, is a requisite of an efficient economy.

The division of labor, in fact, produces no new work. It is merely another way of organizing work, a way that can be more efficient or less efficient, depending on the work that needs to be done. The division of labor—specialization—is characteristic of both highly industrialized societies and the stagnant economies of Asia and Africa, where the same person can be employed to do no work other than fetching water from a well in Mali, or picking tea leaves from a plantation in Sri Lanka. In the U.S.S.R., where specialization is more advanced than in North America, economic stagnation surpasses ours.

But while division of labor can produce no new work, it can prevent new work when the divisions are made along machine lines, when old work is repeated to the exclusion of new work, when the intercourse of ideas is broken, when the diversity that is necessary for the long-term survival of any system is dispensed with for short-term gain.

Division of labor is also a direct source of inefficiency. On an assembly line, all must work to the speed of the slowest worker and the production chain becomes as strong as its weakest link. When the division of labor becomes extreme, as is now the rule, there are long delays as the person in charge of unlocking the office door must be fetched, as stationery must be requisitioned through the office supply department, as purchases over twenty-five dollars need the permission of the manager and those over $100 that of the general manager. Inefficiencies result when uniform manufacturing practices mass produce homes that are overinsulated for some regions and underinsulated for others, or cars that are best suited for neither city nor highway driving.

Division of labor becomes the source of all bureaucracy; quality control over each stage of mass production must be assured; many sets of hands in many postal stations must now handle a letter mailed across a street; the paper burden in triplicate is needed to assure fair play. The slavish observance of chains of command or the transfer of a task from one division of labor to the next wastes more in resources, time, and energy than was ever saved by the main advantage of dividing labor—the elimination of the need to think when performing different tasks.

Division of labor simultaneously becomes a source of efficiency and inefficiency, left as an irrelevant concept when pursued as an end in itself through a blind belief that large-scale projects are necessarily more efficient. Irrelevant except in its potential to destroy the flexibility all cities need to survive.

Under the weight of these inefficiencies, cities become more and more oppressive. In the rush to relieve inefficiency, bureaucracy, and congestion, the very principles that created them—specialization and centralization—are asked to correct them. To control bureaucracy a bureaucracy is created to monitor it; to make cities more efficient diversity is discouraged, to make cities less congested, more roads are added to lead traffic away from the congested area, overlooking the fact that those same roads also lead traffic in.

The range of proposed solutions to urban traffic congestion after the introduction of the automobile covers the full spectrum of centralized thinking. One planner scientifically "proved" that the most effective means of easing traffic flow in congested districts was to rebuild entire areas in identical hexagonal blocks. Another proposed that the entire ground area of the business portion of our cities be given up to the exclusive use of the auto, predicting that "the resulting increase in safety, convenience, and comfort would be so great it would cause the greatest building activity ever witnessed in America." Several practical, double-decked street systems actually appeared (most notably Chicago's Wacker Drive) and plans for triple, quadruple and even six-layered downtown street projects were common.

The head of Harvard University's Traffic Bureau in 1930 suggested we "demolish all of the buildings on entire blocks, use the land not for parks but for open and improved parking areas, with as much benefit from light and air as if they were parks and yet make them pay their way as sound business ventures." A few years later, confident that the car would solve rush hour traffic problems, satisfy all the transportation needs of cities and whatever else was ailing, he proclaimed, "the planners and engineers of today now recognize the full implications of the automotive revolution and their present works give hope for a speedy elimination of many present day maladjustments."

Though the solutions were diverse, the premise was the same: humans don't belong in the center of cities. Working under this premise, all solutions to the urban problem involved either removing the problem or removing people.

When man wanted good housing in the city, good housing was created in the suburbs. When man wanted good working conditions in the city, industries moved out to the suburbs. When man wanted a clean environment in the city, the clean environment was offered in the suburbs.

The assumption underlying man's discomfort in cities was that cities beyond a certain size or complexity or age become unmanageable, unworkable, to be avoided. Yet cities are limited only by how well they use their physical and social resources, not by any immutable natural limits. Today, as two thousand years ago, cities are limited by the amount of water at their disposal, by the efficiency of their sewage systems, by their ability to cope with change. Rome could not expand until its aqueduct system brought water from distant sources. Yet in contrast to the water system to the city, the water system within the city was left in neglect, kept at primitive levels.

Our solution has always been to ignore our internal problems, like the exhaust from automobiles, or to shunt them away, zoning noisy or dirty industries into special areas. We prolong the problems while providing no incentives for their elimination. All pollution, including noise pollution, can be designed away. That pollution is allowed to persist in old and new industrial areas is evidence not of progress in industrial growth but of stagnation in human inventiveness.

No higher form of social or economic organization than the city has ever been known to man, despite centuries of neglect. The city is the natural home of man but as long as man fouls his own surroundings he can't help but be alienated from them. The city is the basis of man's economy; but until the economy of the city can be made to work the economies that depend on the city—whether agricultural or industrial—can't be expected to function well.

We have made the mistake of thinking of the city as an immense iron lung supporting life within, some machine-like creation whose delicate existence is sustained only through massive transfusions of energy and resources. Our temptation has always been either to reject the city for green pastures or to tear down the old to start anew, to build a better machine. We have never dared think of cities as organic wholes, as entities organized around the cycles of nature instead of the rhythm of machines, as complete organisms, capable of self-sufficiency in every respect, able to feed themselves, clothe themselves, provide all their own energy needs, even govern themselves.

We have never dared try to make cities work, to venture back into their centers and regenerate their living tissues.

# 15

## *The New City State*

DURING THE GREAT BLACKOUT of 1965—when a failure at an obscure electricity generating plant in Ontario blacked out all of New York State, most of New England, Pennsylvania, and Southern Ontario—many of the major North American cities were paralyzed. Aside from cars and buses, which do not depend on electricity, almost everything else stopped: elevators could not move people up or subways move them across; lights were extinguished, entertainment was inaccessible, and industries were shut down. Only organic processes were not adversely affected (as evidenced by a boom in babies nine months later). Had oil and gas also been unavailable—cutting off the city from virtually all of its energy supplies—the paralysis would have been near complete. Never before had there been such a graphic example of the city's total dependence on energy. This energy has made possible things good and bad, and that we take it for granted does not diminish its importance to the proper functioning of a city, so dependent on the rapid movement of goods, services, and people.

In response to our energy shortage, energy use in cities has been intensely studied. The conventional wisdom from the conventional authorities claims that cities in the year 2000 will need twice the energy they need now. The conventional wisdom of the most conserver-minded critics counters with studies that show cities twenty years hence can manage on about the same amount of energy as they do now. These conventional wisdoms assume that cities will continue to make conventional mistakes.

We have forgotten that the purpose of a transportation system is to minimize transportation; forgotten that every additional BTU of heat our homes require is a BTU that, until provided, threatens to freeze our pipes or chill our bones. We have forgotten that the scientist's definition of efficiency is work accomplished per unit of energy consumed—the only way to become more efficient in doing any job is to use less energy doing it.

The world suffers from an incredible energy surplus. We confuse gluttony with good taste and equate high levels of inefficiency in energy use with high standards of living. When we look at situations where energy is indispensable for either necessities or conveniences or comforts, the extent to which energy has replaced ingenuity becomes apparent.

We commute from suburb to city because we haven't been able to retain the turn-of-the-century efficiency that allowed us to walk to and from work. We use gas guzzlers because we've been unable to make either the auto efficient or public transit appealing. As a result, the modern urban dweller manages to use 100 times as much energy as his grandfather used getting around cities, consumes more time doing it, and assumes increased risk of injury and death from accidents and exhaust. In fact, there's no reason cities of the future can't run on 5 percent to 10 percent of our present energy consumption: almost all the energy consumed in cities compensates for some mistake, substitutes for innovation, corresponds to a worsening of our quality of life in the same way burning vast amounts of energy made infernos of the cities of the Industrial Revolution.

The wastefulness extends to the land energy consumes.

It takes about one acre to park 200 cars. For a city like Detroit, desperately in need of new efficiencies to aid its rebirth, the boon of its Renaissance Center is burdened by the weight of land wasted in parking—eighty-five acres of parking lot simply to service the Center's employees. But reserve the car mostly for out-of-city uses, use rail instead of road to bring people in and out of cities and—with the savings in expressways and roads, parking lots and driveways—about half the area of most cities can be reclaimed, in effect doubling a city's useful size while moving as many people at greater speeds and lower costs.

Land freed from its need to serve the car can be turned over to a multiplicity of uses, including the provision of enough residential space within cities to enable people once again to have the choice of living near their workplaces. As opposed to roads and parking lots, which bring in little or no revenue to cities, the new residential and commercial space could dramatically increase a city's tax base, allowing it the luxury of providing the lower-cost services available to high-density living . . . precisely the kind of living the preponderance of humanity has always preferred. Despite all our moaning about the dirt, noise, pace, and congestion of the city, man has always preferred to live as near to his fellow man as possible, even in medieval and ancient times when there were no land constraints.

Actually, there have never been any land constraints, only constraints on our ability to use land wisely. It has been this unwise use of cities, not any desire to use cities less, that has led to the depopulation of urban areas. There is no predetermined point at which urban densities become inhuman.

Around the turn of the century, when New York was one of the inspirational centers of the world, 100,000 people per square mile lived in Manhattan (compared to the still extraordinary 65,000 per square mile today). Chicago in the 1920s had a population of 3.7 million living on almost 200 square miles. By 1970, the population of suburbanized Chicago had more than doubled to 7 million but the land used had increased eighteen times. Metropolitan Chicago's density is now about 2,000 people per square mile.

The City of Toronto, built mostly before World War II, still has a density of close to 20,000 people per square mile. Put its new suburbs into that mix, and Metropolitan Toronto's density is less than half that, with a corresponding halving in efficiency.

When population growth is not accompanied by a growth in sanitation services, recreational services, medical services, the city becomes too dense for its population, and people leave for greener pastures. But when cities can increase the density of social and public services, more people can be comfortably accommodated in less city space, and densities can rise. Cities have become more efficient in the past by transporting people underground (saving on roads above ground) or by heating houses with oil or gas (freeing basements for cleaner purposes than coal storage).

With the technologies we now have, there's no reason any city can't have Manhattan's density of the year 1900, with far more comforts, cleanliness, wholesomeness, diversity, options, safety, social services, and ease than has ever been achieved at any time, in any place, before. And if we needed any encouragement to head in this direction, energy shortages will provide them. No efficiency increases are possible without squeezing more work out of our shrinking energy budgets.

Passive solar homes being built today in cold Canadian cities like Regina and Saskatoon are eliminating 80 percent to 90 percent of the year-round energy requirements of conventional homes. These homes sell for no more and look no different than conventional homes, they are far more efficient in collecting the heat that shines through south-facing windows and capturing heat that would normally leave through cracks in the wall. The same insulation that keeps heat inside in winter also keeps heat outside in summer. Similarly, the trees that shade the house in summer shed their leaves when the home needs additional heat in winter. This level of savings may prove to be modest. Experimental homes are eliminating over 95 percent of conventional energy requirements.

Energy-efficient office and apartment buildings have a different energy problem—instead of having to keep the heat in, they must concern themselves primarily with how to keep the heat out. Unlike single family residences there are few exterior walls in big buildings and so few opportunities for heat to escape outside. Conventional buildings, even in the dead of winter, must be cooled off in some areas while being heated in others. The new energy-efficient buildings, such as the Gulf Canada building in Calgary, simply transfer heat from the hot part of the building to the cold, balancing the temperature and eliminating 80 percent of conventional energy requirements at a competitive price. As advanced design techniques develop, large buildings will also be able to eliminate over 95 percent of their traditional energy requirements.

Every new house—every new building—can be built this way. Existing buildings, while they may never be economically redesigned to these standards, will come close. Energy Probe has taken an old, energy-inefficient home in downtown Toronto and eliminated 85 percent of its energy requirements (and Probe feels it could have done better).

With the elimination of most residential and commercial energy needs, and the elimination of most passenger cars from cities, very little energy will now be needed due to the greatly increased efficiency of our urban systems. Cities will be able to tap pipelines transporting fuel from the corn fields and cattle pastures. Cities will be able to plug into the existing electricity grid.

Or cities can provide for their own energy needs.

There is energy in all organic matter; everything that can decompose or be burned has fuel value. Right now we throw all this energy away, yet in our garbage cans and our sewage systems lies the brown gold that will make the city of the near future work.

Every individual produces about 150 pounds of body waste per year, flushed away with 13,000 gallons of fresh water. Unlike Hong Kong, which not only uses these materials but exports them, North American cities treat this waste as a liability that must be processed through sewage treatment plants, leaving behind that residue called sludge. Because industrial poisons get mixed into the sewage, very little is sold as fertilizer in cities, and costs of shipping sludge to rural agricultural areas become prohibitive. The sludge then becomes a waste disposal problem, either fouling city beaches when disposed in bodies of water or fouling land when trucked out of town. Worse still, when in garbage dumps the sludge naturally produces methane gas creating fire hazards.

If the sludge were instead used as an energy source the huge liability it now represents would be turned into a huge asset, producing a non-polluting gas that can be used with the same flexibility as natural gas.

This gas need not be produced at the main sewage treatment plants. It will more often be feasible to have the sewers stop at district plants that can collect the wastes of the neighborhood, or at plants in the very buildings that produce the wastes, whether apartment buildings or office buildings. New generations of composting toilets that draw off methane gas from our body wastes and feed it directly where needed in individual homes will become commonplace.

Until recently, only those countries in short supply of energy have needed to be innovative enough to find efficient use for the energy in waste. In China, for example, 4.3 million of these methane gas plants produce much of the energy needed by industry and individuals. Many of these are relatively small plants operated by communities for fifty or fewer people. It would not take much of a technological breakthrough for the North American engineer to adapt these Third

World designs to New World needs, and produce useful watts from human waste.

Raw sewage aside, as any North American garbage picker knows, over half of the 1,500 pounds that's thrown out each year per person is a paper or a paper product—suitable for burning or recycling. A good deal of the rest consists of table scraps, from which methane gas is also naturally released. Burning either paper or methane directly, however, yields only heat, and modern cities require electricity as well for lighting and running appliances.

Both forms of energy are now possible through the same process, called cogeneration, a technique European industry has been using for decades but North America uses only rarely. When a Canadian or American requires low-temperature heat, he heats water to moderate temperatures to produce his low-temperature steam. A West German or a Dane would take that same water and put it into a high pressure boiler. The high-grade steam that he produces is then run through a turbine, generating electricity. When the steam leaves the turbine it is now the low-temperature steam he wants. Using slightly more of the same fuel, though, he has generated electricity as well and now has two forms of energy for about the price of one.

Until recently, cogenerators have been large and scaled to industry. But Fiat, the Italian automotive giant, and others, have begun mass-producing small cogenerators that run on gasoline, natural gas, methane gas, or alcohol. This technology will allow any individual to be as self-sufficient in his urban home as his wood-chopping rural counterpart managed only with far more effort. But, it will not always be practical, economic, or desirable for energy to be generated at the individual level. Sometimes large buildings, like the Warbasse housing complex in Coney Island or the Park Plaza shopping center in Little Rock, Arkansas, will install their own cogenerators. Sometimes a neighborhood will want central facilities to process all its wastes and be locally self-sufficient.

Cogenerators running on natural gas have already led to greater self-reliance. Rochdale Village in Queens, New York, a 170-acre complex that includes two shopping centers and twenty fourteen-storey apartment buildings, is powered solely by natural gas cogenerating units which provide all its electricity. The waste steam heats the buildings in winter, air conditions them in summer, and provides hot water all year long. Though Rochdale Village does not need to be connected to the electric utility, it is hooked up to Brooklyn Union Gas Co. for its natural gas. With a different design in plumbing, Rochdale could generate its own supplies of methane gas to fuel the same cogenerators that are now burning purchased fuel.

Once energy independence is proclaimed at the local level, former economic disadvantages will become the basis for new efficiences: new work will be fashioned from old.

Individual households won't always want to process their own wastes, preferring instead to sell them to local entrepreneurs who will make it their business to pick up home fuel as oil trucks now deliver it. Some entrepreneurs may run community gas and electric utilities; or some may run local greenhouses, using the wastes for both fuel and fertilizer. In the late 1970s, organic fertilizers became cheaper to use than chemical fertilizers, convincing farmers to start collecting cow manure again. Urban entrepreneurs will have the advantage of human wastes pre-collected, and often pre-separated into their methane gas and sludge components.

Producers of large amounts of garbage, such as restaurants and schools, may start businesses on the side to take advantage of their windfall. And large consumers of energy, such as public transit systems or factories, will benefit from assured, local supplies.

Energy, however, will be available from a wide variety of urban sources. Solar collectors can trap heat and solar cells can generate electricity directly while windmills can be placed on any roof or attached to any wall. Wherever the wind does not blow strongly enough it can be amplified by channeling wind through corridors that are no more complex to construct than a brick wall or a funnel. These wind tunnels are often built inadvertently by sloppy architects. With a little forethought, wind tunnels can be incorporated as a design feature of all construction projects, providing at no additional cost high quality forms of non-polluting energy that can be tapped whenever needed.

In an optimally designed urban environment the liquid and solid wastes we generate alone can produce all the electricity and several times the heat we'll need. That leaves us with an immense heat surplus. The sewage and table scraps that produced methane at individual, large-building or community-scale plants have become rich sludge and compost, made even richer through the fermentation process that released the methane gas. Because the sewage stayed in local systems, segregated from all the industrial wastes in the centralized systems of cities, this compost is now free of pollutants, a valuable form of fertilizer available in quantities great enough to ship out to argricultural areas. But it will often make more sense to keep this compost in the city where, with the excess heat available, it can be used to grow the city's fruits and vegetables.

Despite our notions of the limited space in cities, the area of land in any city that could be made available for agricultural purposes exceeds the land area of the city. This is best understood when a city is viewed from the air, and the land area is seen to be the sum of the roads, lawns, parks, roofs and tiers of balconies of the city. With most of the roofs flat and most of the roads reclaimed from the automobile, as much of the city can be converted to farmland as its residents desire.

In a typical city like Chicago the agricultural potential is staggering. It takes about 250 square feet of intensively farmed land to provide the fruits and vegetables an individual needs over a year (less, if highly intensive farming practices are used). With a roof garden atop every building, lawns all converted to growing produce, and roads resembling an agricultural cornucopia, each person in Metropolitan Chicago could grow enough to feed himself and a community of fifty on the farm. Chicago, in other words, could feed the state of Illinois many times over if it devoted all its potentially available land area to farming; it could feed itself alone by tilling only 2 percent of its land area.

In addition to the land area visible from the air, there is space on window sills, in home greenhouses, even in basements, caves and old abandoned mines.

The nickel mines around which Sudbury, Ontario, was built, are now being seen as a source of fresh vegetables, an underground idea that goes back to at least the time of Louis XIV when the caves of Paris (created by quarrying stone to build the city) were used as a logical place to grow the city's mushrooms. (Mushrooms like the dark, damp and chill.) By 1867, the year of Canadian Confederation, some of these caves extended underground for twenty miles and produced 3,000 pounds of mushrooms per year. For the urbanite without a cave, add water to a plot in the house basement and you can expect four pounds of mushrooms per square foot.

The liability of unused, often unusable basement space thus becomes an asset, with its cold put to good use. Similarly, the heat atop buildings—they're usually about five degrees hotter than ground level, prolonging the growing season—is an added bonus to the productive use of wasted rooflands, while excess heat from the buildings in winter can more than support year-round commercial greenhouse industries. To return the favor, rooftop greenhouses help to insulate buildings, keeping them warmer in winter and cooler in summer.

The same degree differential that lengthens the growing season of roofs over roads lengthens the growing season of city over countryside. Unsheltered from the chill of strong winds, not benefitting from a hubbub of activity over, under, and on its surface, rural areas provide a less hospitable agricultural environment than city spaces, scarce though city space may seem to be.

A surprisingly large number of North Americans are already involved in cultivating their basements, window sills, whatever land they can get. According to a Gallup Poll, 7 million people without access to land would garden if they had access to land, while about 35 million households, or 43 percent of all families, already grow their own fruits and vegetables. This despite the lead pollution in cities from cars, and despite the often inconvenient location of the few garden plots made available by cities.

City-sponsored programs serve thousands of people in Boston, Vancouver, Chicago, Toronto, Detroit, Los Angeles, and elsewhere while Pennsylvania, Connecticut, and Massachusetts have statewide community gardening programs. Schools, too, have gotten green thumbs, involving over 20,000 children in Cleveland and about 400 schools in Alabama.

This non-farm produce has become a potent economic force with the value of home-grown vegetables estimated at over $15 billion dollars continentwide, and the economics of agriculture dictate that more and more urban land will be cultivated as rising energy prices can't help but inflate food prices. About 15 percent of the total energy we use goes into the process of growing food and getting it to the tables, and close to 40 percent of this energy use—the energy that goes into farm production and the energy that goes into the transportation and distribution of food—is eliminated simply by growing food in the city.

But economics will not be the only factor. While most of those who turned to urban agriculture in 1974 gave the cost of food as their main motivation, by 1977 most felt that the recreational value of gardening had become more important to them than the money saved (an average of $375 that year on each gardener's food bill). Should nutritional value and taste become a factor in the North American diet, the non-economic impetus toward city farming will further increase.

More than fruits and vegetables can be grown in cities—in fact, in our major cities today chickens, rabbits, goats, and bees are being raised, producing eggs, dairy products, and honey at a price and freshness that's hard to match. A city-made egg, for example, would surprise most city dwellers, not because it can be produced at one-third to one-half of the cost, but because its vivid color and true taste clash with all their years of experience eating six-week old supermarket eggs.

Chickens, with one of the toughest digestive systems in the animal kingdom, are ideal big-city birds. They will eat almost anything in the nature of a table scrap, needing only four to five ounces of food and an eight-ounce glass of water per day. They also need very little space—four square feet per bird—and very little attention. For two years one Vancouver family, the Mckinnons, found five eggs each day during the summer (tapering off to two during the winter) under their five laying hens, which, Mrs. Mckinnon claims, were less trouble to care for combined than her one cat.

When Vancouver City Council found out about her backyard coop they invoked bylaw number 4387 which outlaws chickens in the city. The bylaw exists, said the city health department, because chickens attract rats and disease, have a foul smell, and produce unwelcome noises. Yet the Mckinnons have never had any problem with disease and have seen no sign of mice, let alone rats. When Mrs. Mckinnon took a petition around to each of her neighbors, none registered any objections, most said they enjoyed what little they had heard of the chickens, as it reminded them of the country, and many weren't even aware she kept chickens at all.

City council was not impressed. Although Vancouver manages to keep 25,000 dogs under control, although a chicken makes less noise, in volume or persistence, than a car, a dog, or a lawnmower, although the city tolerates innumerable industrial noises and smells, although snakes and reptiles are considered OK, city council squawked at the Mckinnons' chickens, and forbade them to continue raising their hens.

Similar prohibitions prevent city dwellers from having a cheap and wholesome source of a wide variety of fresh meats and dairy products, in the belief animals have no place in the metropolis. They forget that the exclusion of farm animals from cities is a relatively recent phenomenon. A 1906 census of American cities showed there to be one chicken for every two urban inhabitants; it has only been since World War II that mechanization has taken the chicken out of the hands of the individual poultry raiser.

Though it may be demonstrably clear that chickens do no city harm, to rescind the anti-poultry bylaw, city councillors think, will be a step backward, one hurting the forward image of their municipalities. Ironically, two of the cities considered the most progressive in North America—Toronto and Berkeley—have not legislated their hens out of town.

Families like the Mckinnons who enjoy farming are more the rule than the exception. Next to eating, sleeping, and reproducing there are few human activities involving as many North Americans. That gardening is so pervasive, despite all the roadblocks, speaks well for the potential of what cities and city dwellers would naturally do were it not for the barriers put in their way. Without the wasted space taken up by unnecessary roads in cities, urban farmland could be located so close to everyone that—whether city farmers were motivated by profit or recreation—cities would increase their supply of fresh food while reducing the cost of eating. Without the unnecessary cars on those unnecessary roads, pollution levels would drop to the point that food grown at road level would become safe to eat. Without restrictive bylaws, which prevent cities from producing their own meats and dairy products and prevent producers from selling even their fruits and vegetables to local shops, urban entrepreneurs would find that their hobbies turned a profit. Without the huge subsidies spent on keeping energy prices low so that food can be produced inefficiently, then shipped inefficiently distances of 3,000 miles and more, cities like Boston and Montreal would grow their own lettuce instead of importing California produce. Without those same subsidies allowing our energy-production systems to stay so inefficient, we would be thinking of our energy supplies as part of a closed system and benefitting from the heat we presently release into the air and the garbage we presently dispose of under the ground.

If we think of the city economy as a closed system and radically reorganize the structure of our production machine to resemble an organic mechanism that recycles its own wastes and responds to seasonal variations, the agricultural needs of city populations can be met with uncanny efficiency. In winter, when we don't need as much fertilizer, we can burn our wastes for heat. In summer when we don't need as much heat, we can use our wastes as compost.

To supply cities with the water necessary for agricultural production, we need only recycle the 13,000 gallons per year of fresh water each of us now flushes away with our wastes. Instead of accepting the overproduction of heat as a necessary inefficiency in all our energy uses we can use that heat to warm greenhouses for vegetable production. And where inordinately large amounts of waste heat are available as will happen in summer, the greenhouses can provide us with bananas, pineapples, dates, mangoes, and papayas.

Recognizing that there are efficiencies to be gained by hopping on nature's bandwagon applies to our inorganic resources as well. Cities, in fact, need not import all their raw resources. They can be their own mines, their own forests. Man's accumulated resources of the last few millenia—the sum of all his mining and quarrying activities—are now concentrated in the cities of the world, available for use and reuse. We only have to form a closed loop with all the city's resources.

About 38 percent of the garbage we throw out is paper. Although it will often make sense to burn it as fuel, more often it will be considered too valuable a resource for incineration, saved for recycling as soon as sensible paper recycling systems are in place.

The metals, plastics, glass and other inorganic components of garbage (comprising 22 percent) will also be considered too valuable to dispose of casually. What will provoke our second thoughts will be the provocatively high price of energy. Recycling aluminum saves 95 percent of the energy needed to produce aluminum from bauxite; recycling steel saves 74 percent of the energy needed to produce steel from iron ore. These energy savings are available because scrap is already processed material, incorporating the energy expended in the mining, milling, and refining of raw resources. For this reason, there are also significant energy savings when paper, rubber, or virtually any other material is recycled.

Unlike mines and paper mills, which are difficult to relocate close to both their source of supply and their customers, metal or paper recycling plants can be located wherever they're needed, minimizing the transportation expenses necessary in shipping minerals from distant mining towns to metropolitan areas, and maximizing the use of second-hand resources that would otherwise be dumped in a landfill at public expense. Metals can also be mined before they become household garbage.

With human sewage collected locally and removed from industrial wastes, our municipal sewage systems will now be highly concentrated with mineral wastes from various industries, rich in metals, minerals, and petrochemicals or their equivalents. Technologies are now available which economically remove trace elements from a waste stream at concentrations of several parts per billion. But if these technologies were applied at the stream's various sources, before the wastes of one industry get mixed with the wastes of another, the economy of separating their metals would become that much more compelling.

Technologies that recycle with close to 100 percent efficiency still have limited applications, but even at recycling's present pace, economically converting 95 percent of the scrap into good-as-new material, the present store of resources within cities can supply them indefinitely. The small resource deficit each time a good is recycled can be more than made up for by a small infusion of human ingenuity. Although a recycling system is more efficient than one that dumps resources, it compares poorly to a system that minimizes recycling through reuse.

Glass for example, can be recycled easily enough, but not until it's reused does it begin to live up to its potential. A refillable pop bottle averages twenty-eight lives compared to the premature death of the disposable. Additional resource efficiencies (and economic savings) can be found when bottles are standardized. In places like Alberta, where most beer bottles are standardized, the return rate is 98 percent. There, guzzlers must rely on the artwork on the label and the taste of the beer to distinguish between brands, unaided, as New Englanders are, by the shape of the bottle.

The majority of ketchup bottles, wine bottles, and jam jars can be standardized to the few dozen functional designs consumers desire rather than the thousands of cosmetic shapes marketers dream up. With few exceptions (such as liquor, for which decorative bottles are prized) consumers would be no more willing to pay extra for oddly shaped glass containers than for decorative tin cans. (Standardized containers, like standardized sizes for nuts and bolts or standardized automobile parts, do not result in standardized products unless the container itself becomes the product. Depending on its function, standardization can be either an aid or an impediment to diversity.)

Costs can be kept down and recycling can be kept to a minimum, sparing not only glass resources but over 95 percent of the energy required by the glass-making industry. Ironically, the switch from glass-making to glass-reusing creates more jobs in the bottle-cleaning services than are lost in the bottle-making industry and even the 5 percent of energy still used is cheaper—the high-quality energy needed in glass making can be replaced with excess waste steam made available from our cogeneration units.

In the manufacture of virtually every product, the lengthening of a resource's life can be accomplished by merely making the product more durable and planned obsolescence less acceptable.

For the first time in history, a modern industrial strategy for cities can now be formed based on the most efficient utilization of all the city's resources rather than a regional or national industrial strategy based on the mutual exploitation of different areas for the net benefit of neither. Regional differences will still exist, but they will show up as differences in urban industrial strategies based on the differing cultures and differing climates and geographies of different cities. They will be determined increasingly not by the whim of some central government that moves national resources around as if on a chessboard, but through urban strategies designed to meet local needs.

People in one city, in one neighborhood, will have very different ideas of how to live than people in neighboring cities or in the next neighborhood. Some neighborhoods will object to city farming; some even to corner stores. Some will want commerce to boost community morale; others to fatten their pocketbooks. Instead of balancing the demands of one area with the demands of another through compromise where, inevitably, both sides lose, individual areas will suit themselves and benefit from the diversity available when other areas do likewise.

Cities can be self-sufficient in food, even exporters of food. They can be self-sufficient in energy. They can continue to be exporters of manufactured goods, using the manufactured imports from other cities to provide them with their future raw resources. But it will not be important to do these things—to be 100 percent self-sufficient—unless we expect another ordeal like Leningrad's 900-day siege. The importance lies in retaining these options to equip cities for the certainty of a future whose unknown factors will alter our most basic assumptions of what we'll need, what we'll value.

The way society is currently structured, this flexibility does not exist: our options are limited by our failure to emulate the inherent stability in nature. If nature has taught us anything, it is that strength lies in diversity, seeds of destruction in centralized systems because of their inability to adapt.

It is this inability to respond to change that has led us to our present precarious economic state, not the ill will of powerful individuals, not the corruption of power and money we attribute to our enemies.

This inflexibility is a function of size and structure. It is a feature of all large institutions, and all monopolies. It is only this inflexibility that has made the energy crisis of international scale, and now that the energy crisis exists it is only this inflexibility that is preventing local solutions to what could be local problems.

# 16

## *Beyond the Barricades*

THE CITY OF Toronto sells the old newspapers that residents leave for pickup each Wednesday, producing a profit for the city while saving on garbage disposal costs. The newspapers are bought by paper recyclers, providing them with a resource they need to meet their production commitments. When the recyclers need more paper than the city of Toronto can provide, they import old newspapers from Buffalo, New York, ninety miles away. Making new paper products from old is so profitable (up to four times cheaper than using new pulp) that it justifies the high shipping costs involved in moving goods across borders.

Yet York, a large suburb bordering Toronto, has stopped collecting newspapers for recycling. It found the recycling business unprofitable, and instead pays to have newspapers hauled to a landfill dump along with the rest of its garbage.

Inexplicable? Not at all. Impossible in a profit-motivated society? Ironically, this situation exists because of the profit motive. Some aberration that is peculiar to Toronto and doesn't mean much in itself? This "aberration" and others like it have now become the norm, belonging to a new class of phenomena known as "institutional barriers." An institutional barrier (which we intuitively blame on "the system") typically makes us shake our heads in dismay, forever surprising us with its apparently inexplicable stupidity. It is at work in corporate departments unnecessarily spending all their allocated money before the fiscal year expires. It may cause a labor union to resist automation, despite raises, retraining, and full guarantees that no worker will be forced out of an existing job. The practice of including electricity in the price of the rent is an institutional barrier that makes a non-paying proposition of turning off unneeded lights. The effort to bring glass to the recycler may be counter-productive as more energy is expended in getting the glass to the recycling depot than is saved through recycling. In each case, an institutional barrier is at work, seemingly playing havoc with our desires, and our reasons. Yet the institutional barrier is actually part of an orderly process that perfectly meets the objectives of the present system.

If the corporate department did not spend its money in time, justifying an increased budget would be difficult, even if the increase was really needed—that's the way accounting procedures in large institutions have often been set up to guard against greater wastes. If the labor union agreed to terms favorable to its workers but resulting in fewer union members, the result could be the demise of the union, and a weakening of its future bargaining power (not to mention the job security of the union officials). If electrical utilities had to bill every apartment separately, their billing costs would increase while revenues from sales of electricity would decline—inconveniences no monopolies have to put up with.

But although institutional barriers make sense for the system, they're detrimental to society. In every case, irresponsibility has been encouraged through some form of bureaucratic shortcut or bias toward the *status quo*. In every case, the large size of the institution determined the barrier's effectiveness: had the barriers been erected by small institutions, their effect would have been negligible, impossible to enforce, and so, short-lived.

These barriers do not require the ill will of individuals. They are a necessary consequence of any large institution that carries out its mandate well but without needing to recognize the costs to others. For the more it promotes its particular needs, the more it can impede the progress of others, evolving systems that optimize society's parts at the expense of the whole.

The same paper company that recycles Toronto's old newspapers also makes paper products out of virgin wood, felling the trees, running the saw mills, and breaking the wood fibers down into pulp as part of the paper-making process. Though profitable, recycling old paper represents a minor part of its multinational operation. The company has heavy investments in all aspects of its operations, and can ill afford to keep the bulk of its machinery idle. For this reason it produces most of its paper from virgin wood. Because it regularly needs more pulp than its lumberjacks can provide, the paper mill supplements its own supplies by buying the paper collected by the City of Toronto. But it doesn't need enough to also ensure a market for York's paper—only when the demand is especially great will it buy paper from York, Buffalo, or wherever else newspapers happen to be collected.

The paper mill could produce cheaper products by depending on recycled stock for normal operations and using virgin wood as the supplement. But while one end of the paper mill's operations would benefit, the other end would lose, and overall profits would decline. The mill only buys paper at its convenience, optimizing its own operations but triggering the misallocation of thousands of tons of valuable resources. In doing so, its motive is not harm to society but profit for its shareholders. It is acting rationally from its own perspective. The barriers created to recycling are a function of the way the pulp and paper industry in Ontario works as an institution—it is organized as a monopoly.

The same barriers to recycling recur in the paper industry across North America. Other corporate institutional barriers prevent glass recycling, metal recycling, and recycling of garbage into fuel. Institutional barriers within municipal governments also prevent more recycling of paper. The cost of collection of paper and the costs of disposal of garbage are often calculated in separate sets of books, with the department in charge of each interested only in keeping its own costs down. If it costs forty dollars a ton for one department to collect paper which fetches thirty dollars a ton for another department when sold to a recycler, and twenty-five dollars a ton to send the paper directly to a dump, which gets nothing back, the latter course is often taken. Only in large institutions, where bureaucratic procedures are needed to maintain control, could barriers of this nature stay in place long after the wisdom of another course is clear.

This same inflexibility in large institutions prevents entirely new garbage pickup systems from being created. In the same trip not only paper but glass and metals could be collected for recyclers, tripling the efficiency merely by keeping garbage segregated. The energy savings from recycling glass would now become significant, and the individual efforts made to separate glass would truly be fruitful. These institutional barriers—unnecessary intrusions into the market mechanism—are all that prevent systems from adapting to change.

Subsidies represent the greatest institutional barriers. Removing subsidies alone would result in profound changes. For example, without the subsidies on transportation, we would make less and less use of the automobile as people moved closer to their workplaces and switched to public transit. With their streets decongested, municipalities will realize that road space is available to implement ideas they have been toying with for a decade and more—exclusive bus lanes, exclusive transit ways, pedestrian malls, and bicycle paths.

At this point, the city may have already reached a point of no return, having gathered enough momentum to break through the other institutional barriers between it and the organic city state.

One breakthrough will be the bicycle, one of the most underrated North American vehicles, yet far more a factor in some of our present transportation systems than is realized. In the city of Toronto, for example, 140,000 one-way bicycle trips are made by residents on an average daily basis, 88,000 of which are not primarily recreational in themselves, but serve as transportation to work-or-play destinations. Despite fear of being hit by cars, the unavailability of secure parking, and concern over pollution (bicycling for one hour in the downtown core is the equivalent of smoking a pack of cigarettes), and even counting those too young, too old, or too ill to bicycle, almost one-third of the population shows its attachment to the bicycle by cycling at least once a year. On the whole, Toronto residents bicycle half as often as they take public transit.

The average cyclist is thirty-nine years old (if male) or forty years old (if female), is the head of a family, and generally has an average or above-average income and education. The energy consumption of a bicycle is fifteen calories per mile, an equivalent of 1,500 miles to the gallon. More than half of all bicycling activity is by over sixteen-year-olds, saving, over a year, 100,000 barrels of oil in Toronto alone, had their trips been made by car. By European standards, these figures are far from astounding. In Copenhagen, Denmark, 17 percent of all trips are by bicycle, in Upsala, Sweden, 20 percent, and in Rotterdam, Holland, 43 percent, helping to explain in part the great efficiency of these affluent Western European cities. By North American standards, Toronto is an anomaly. A study for the

Bicycle Institute of America indicates only 1 percent to 2 percent of all commuting trips in the United States will ever be by bicycle.

But without the North American institutional barriers to cycling, and with some of the European incentives—good bicycle path networks, shorter distances due to less suburbanization, and true fuel costs for public and private transportation—the bicycle can compete well with motorized transportation while offering side benefits in recreation so valued by the continent's hordes of joggers.

Eight-lane city streets could become six-lane streets, six-lane streets four, four-lane two, and two-lane roads could become one-lane, one-way streets, reversing the traditional road-widening ritual (as has already begun in Victoria, B.C.). At first, the streets would retain their transportation function, turning more of themselves over to public transit, bikeways, and service roads for goods and emergency vehicles (as has already happened in Salem, Massachusetts, Philadelphia, Pennsylvania, Portland, Oregon). But as the car becomes outmoded in city cores, with fifty at a time being replaced by a single streetcar, and both car and streetcar not necessary when more people walk to work, roads—still 50 percent of the land area of cities—will become largely outmoded as well, and turned over to other uses.

Restaurants will spill out into the city streets in the form of sidewalk cafes and kiosks. Malls—already tried and usually proved in almost every major North American city—will reclaim not just one urban artery (Ottawa), but whole sections of town (Vancouver, San Antonio). Six- and eight-lane urban roads will be wide enough to permit the building of offices and apartments. Narrower roads might have low-rise buildings bridging both sides of the street (Toronto is even considering bridging an expressway with housing), leaving enough space for nighttime service vehicles, with above-ground windmill complexes built to take advantage of architecturally created second-storey wind tunnels.

The ground floors of buildings can be dedicated to all those browsing activities which so attract humans, filling windows with fantasies of clothing, literature, photographic services, and floral arrangements—while the upper floors can provide the housing. Of course, this trend to over-the-store apartment living is already firmly re-establishing itself in cities across the continent, with high-speed elevators replacing the walkup.

Roads will suddenly become revenue-producing properties for cities, just as cities have discovered they can boost low taxes received from parking lots by building instead. The majority of all roads, though, can be devoted to the commercial activity of the mall and the recreational and agricultural potential of reclaimed green spaces.

The cost of converting sewer systems in existing buildings will often be prohibitive. But for the new complexes, fuel costs will be prohibitive if they don't utilize their sewage wastes, yielding both methane gas and a rich, odor-free fertilizer. Some of this fertilizer may be available to building residents, who may also be offered plots on roof gardens or surrounding land for a small supplement to the rent, the way parking spaces are currently allotted. The rest of the fertilizer can be sold to local entrepreneurs or exported out of the city.

By coincidence, the very fruits, vegetables, and foods which most lend themselves to urban exploitation (lettuce, tomatoes, eggs), also are the greatest energy consumers, generally needing an investment of ten calories of energy for every calorie of food produced. Growing these inside cities, with "free," locally produced fertilizer and waste heat, would minimize, or negate, the price hikes these products would otherwise face with increasing energy costs.

Grains, and other field crops, on the other hand, which do not lend themselves as much to urban agriculture, are the most energy-efficient foods to produce, needing an investment of only one calorie to grow eleven calories of food. But the need to cut down on transportation costs will nevertheless keep many out-of-city farming operations close by, with grain fields and animal farms out of scent but not out of mind. Until the 1960s, eleven pig farms on the outskirts of Toronto benefitted from the uneaten, chef-prepared foods of the city's restaurants. Only when suburbanization forced pigs far from the urban center did society need to allocate them special feedstocks while feeding restaurant food to the local dump. It will not be hard to once again shorten the distance between city and farm.

The feasibility of meeting urban food requirements is illustrated by China, where, without the benefit of western agricultural technology, 80 percent of the vegetables consumed in cities is grown within ten kilometers of them. Or by southern Germany, where virtually every available square foot of land in and around cities is treated as immaculately manicured farmland, economically growing a few rows of asparagus next to a few rows of potatoes next to a few rows of some other vegetable.

There may be only one comparable North American example—the Chinese-run farms in the Big Bend area near Vancouver, British Columbia. The total area under cultivation amounts to only 187 acres, yet this tiny farming district produces nearly all the carrots, celery, radishes, green onions, parsley, spinach and fancy lettuce for the entire province. Chinese farms average $11,000 worth of vegetable produce per acre, or almost ten times the $1,300 per acre return their highly mechanized farming neighbors receive.

The difference lies in the manual efficiencies achieved by small-scale operations. No farm exceeds twelve acres and some cover no more land than an oversized suburban backyard. The largest piece of machinery to be found is a tractor. Most work is done by hand, with wheelbarrows, hand ploughs, and hoes the mainstays.

Because the farming is on a manageable scale, highly intensive agriculture can be practiced. No land is wasted; the vegetables are planted as closely together as space allows, and wide rows don't need to be dedicated to the passage of planting or harvesting equipment. For eight months of the year, fast-maturing cash crops fill the fields. As soon as one crop is harvested, the ground is prepared for another sowing. The land is never left empty for more than a few hours during growing season, and most of the Chinese farmers keep crops in the field until November.

As the vegetables mature, farmers of this multi-million dollar business pick, wash, and pack them into boxes, all by hand. The crates of freshly picked vegetables are then trucked to their nearby urban markets.

The rate at which we turn to urban agriculture will ultimately depend on how society decides to value urban agricultural land—as a parklike amenity or a commercial property to be taxed as any other. Land values aside, small-scale, organic farming has already proven itself as profitable as large-scale, conventional farming (and rising energy prices will soon overtake the perceived advantages of conventional farming).

Urban land that is not now used will be the first to be converted to agricultural uses. In Seattle, power line rights-of-way are being groomed to grow black cottonwood and red alger—trees as energy crops that can be brought to maturity in four to six years. In Hartford, Connecticut, the land is available right in the city core—vast tracts of it left behind by industries in their stampede to the suburbs and the sunbelt. Also left behind was an unemployment problem especially severe for youths. Treating the unused land and unemployed youths as usable, employable resources, Hartford put the two together to produce food that is now sold twice weekly at the Hartford farmers' market. An expansion of the program is expected to lead to the creation of three more farmers' markets and spin-off industries, such as community canneries. Solar greenhouses are also in the works.

Hartford's leadership in urban agriculture is no random event: because of its location in New England, far from food production centers and needing to import 85 percent of its food, Hartford's food bills have been 6 percent to 10 percent higher than the U.S. average.

Whether produced on former industrial land, rights-of-way, rooftops, or roads, the home-grown produce will find its way to local grocery stores and restaurants with farmers' markets flourishing alongside, providing consumers with a diversity of ways to buy their fresh produce.

Urban inventors will once more turn their minds to improving agricultural processes; only this time they will adapt them to urban use. They will devise self-watering and self-fertilizing mechanisms, develop more intensive farming practices, find innovative ways to preserve agricultural wastes as future energy and composting resources. And the raw resources that will be needed for the new urban industries can come from the city itself: the soil that will reclaim the streets will be a byproduct of methane production; the glass for the greenhouses can come from glass recycling operations; the metal for the new windmills and agricultural implements from metal-recycling plants.

New work can then continue to be refashioned from old: generation after generation of advanced cogenerators and other energy systems; new public transit vehicles and delivery systems adapted to rail; advanced communications systems to reduce transportation needs; an entirely new range of social services designed for pedestrian cities to provide, at far lower cost, equal or better facilities than presently available. The changes will affect all our institutions, all our activities, all our perceptions, in direct relation to their energy dependence.

Not surprisingly, the police force—now heavily dependent on the automobile—would be one of the first institutions to change. In a city free of automobiles, the criminal will have to devise new ways of making his getaway. Not needing to keep up with criminals in cars, police departments will be able to trade in their vehicles for walkie-talkies and shoe leather, and offer superior protection by being part of the working environment, in a better position to mediate problems or notice suspicious goings-on in the neighborhood. Society saves (perhaps) in less crime and (surely) in lower policing costs.

Society can also save by locating police in many small storefront neighborhood precincts instead of expensively large, imposing police stations, integrating the police into society to make the cop-on-the beat once more as approachable as ice-cream vendors or dentists, to be seen, not through car windows, but in the flesh.

Small-scale advantages in medicine of the kind being found today in community health clinics can become contagious, with first aid clinics and emergency wards locally available to provide faster medical attention at less expense to the sick. As our population becomes increasingly aged, the additional mobility provided by improved public transit and improved availability of goods and services will relieve pressure for large institutions to service the aged, rendering obsolete new social services that have become necessary in our centralized cities, such as volunteers to take senior citizens to the shopping center and libraries-on-wheels.

Small-scale schools, too, such as the ones prospering in Peel County, Ontario, can proliferate, producing better students through more personal relationships with teachers at a lower cost per pupil than is managed in the larger schools at the same school system.

Before the eighteenth and nineteenth century cleanup of our cities, living above our commercial and social facilities, or even mixing them, was not uncommon. Shopkeepers used to rim churches with their stalls, a seemingly irreverent use of church land for most of those who have not seen the revival of this medieval practice in modern Munich, rebuilt after the war without many of the mistakes of the intervening three centuries.

Without institutional barriers, our reshaped cities will be dominated not by large social and business institutions but by community social services and local entrepreneurs, with a personal stake in making their communities work.

The striking contrast between large and small corporate forms can best be seen in Sudbury, Ontario, whose economy was totally dominated for most of this century by INCO, the multinational mining company. As soon as it became more profitable to extract another form of nickel ore, INCO began shifting its operations away from Sudbury to Indonesia and Guatemala. Sooner or later the town of Sudbury faced bankruptcy, slated to become one of the grandest ghost towns in Canadian history. What may ultimately rescue the town is a drive toward self-help, toward diversificiation of industry. Spearheaded by a group of entrepreneurs known as Sudbury 2001 (a group scoffed at by both the multinational corporate giant and its multinational union, the Steelworkers, but with great local support), Sudbury's aim is to find thousands of ways of supporting itself instead of relying, almost entirely, on one industry parachuted in from outside. Ownership will then tend to be local, by people with a direct stake in their community who are in the best position to know what local markets can be created, and what business opportunities are available.

To do this, Sudbury will need more means of support under its own control. As things stand, at every turn Sudbury is being hamstrung by the Ontario government whose well-meaning subsidies do nothing to resolve the structural problems of the local economy. The commercial greenhouse that will eventually provide 10 percent of the region's fruits and vegetables, the solar panel manufacturing concern, even the mohair industry being tried in the Sudbury basin, all suffer from institutional roadblocks placed in their way by higher levels of government.

The inherent inefficiency of having to go through senior government bureaucracies to get things done at the local level will lead to increasing pressure for state and provincial governments to return authority to municipal governments. Metropolitan governments will either assume less control over their member municipalities, or disappear altogether. At least one city—Fairfax, Virginia—has been forced to opt out of the bus segment of the Washington Metropolitan Area Transit Association (METRO) when faced with a projected threefold increase in Metrobus costs, and the City of Westmount, in self-defense, has decided to go back to policing itself after years of being taken for a ride by Montreal's Metropolitan police force.

Since 1972, when Montreal Island's police forces amalgamated, traffic violations and break-ins in Westmount have tripled. Rising crime rates have also forced many of Montreal's twenty-eight suburbs to set up private police forces. Montreal's highly centralized force is a failure at more than providing police protection—at $200 million per year, it is the costliest in the country and an increasing source of irritation to its member municipalities.

These are not isolated grievances. In a survey of chief elected officials in a large Northeastern U.S. region (including New York, New Jersey, and Connecticut), only 60 percent felt that regional governments might someday benefit their own localities, and almost 80 percent reported frequent conflicts between regionalism and home rule.

In Canada, the Federation of Canadian Municipalities has been formally demanding that "municipalities be recognized as a distinct level of government under [a] new constitution" since 1978. By mid-1980 a primary concern of Canadian municipalities had become autonomy, a view forcefully paralleled in the United States where Boston Mayor Kevin White termed the city's "economic subservience and dependency . . . just cause for revolution in the Boston Tea Party Tradition . . . We must insist on greater sovereignty for cities, greater mastery of our own affairs, a louder voice in determining our future."

Institutional barriers are as old as institutions, but not until this century—with the growth in monopolies and growth in government—have they acquired such significance that an entire new discipline was developed to overcome them. This discipline is called planning—a process designed to minimize the effects of human frailties. By breaking problems down into various components, planning replaces our vague, qualitative intuitions about where we're heading with a series of vague, quantitative intuitions now elevated to a science.

Unavoidably, planning has become one new layer of bureaucracy, and thus an institutional barrier of its own.

# 17
## *Shoot the Planners*

DEMOCRACY IS A stupid system, or so any visitor to the USSR is advised. Instead of taking the necessary measures, politicians in Western democracies bicker, sometimes for years, over insignificant details. Unions are allowed to go on strike, materials are shipped not to where the state feels they are most needed, but to where they fetch the best price. Special-interest groups acting in their own selfish interests can prevent great works from being accomplished for the greater good of all. Leaders are weak; everything is accomplished the hard way.

Of course the Russian sentiment is correct. The contrast between their economic system and our market-oriented economic system is striking. Just look at the difficulties that would be involved in getting a free market to accomplish a task as simple as producing food in our cities.

Before we can grow food in cities, we need a poison-free environment so that the food is edible. Before we can have a poison-free environment, the car must be removed from urban centers. Before the car can be removed from urban centers, alternative transportation systems must be made available. For these transportation systems to be developed, society must be able to afford them. For society to afford public transit, transit use must be increased. Before use of transit can increase, the costs of transit's competition, the car, must be recognized. For the full costs of fuel and the social costs of the automobile to be felt, the subsidies must be removed from private transit. For the subsidies to be removed from the car, the highway lobby and oil lobby need to be diminished. For these two lobbies to be weakened, two of the biggest monopolies in the history of the world must be broken.

Of course, more than clean air is required to grow food in cities. Non-toxic fertilizer must be produced, land must be made available and . . . one of the most formidable obstacles . . . the human being . . . must be persuaded that all this is a good idea. This accumulation of obstacles to good works appears not only insurmountable, but socially unjustifiable, especially for the bureaucrat with vision, who knows exactly where society needs to be, and wants to waste no time getting there.

Much more appealing than the convoluted market approach is to simply legislate whatever is needed. Pass a law setting aside a certain percentage of a city for agriculture. Legislate the car out of growing zones, except during times of emergency and for deliveries, when vehicles can be monitored to keep exhaust emissions within permissible limits. Create a food-marketing board to ensure no unfair competition from food grown in outlying areas, and provide subsidies to the urban farmers to offset the subsidies the rural farmers get. Now the impatient bureaucrat no longer needs to wait for results. Large systems can be created full blown to accomplish large results in relatively little time.

This method, called planning, has such obvious advantages that planning departments full of planners are now a feature of virtually every bureaucracy in the world. All levels of government depend on them. Universities now train planners and graduate people with degrees in planning. And planning has broadened its horizons, to include urban planning, economic planning, energy planning, and social planning. Planning has become absolutely irresistible, and planners are quick to point out the unrealized potential thwarted by short-sighted politicians making poor decisions for fear of losing votes.

Occasionally, the zeal for reform has coincided with the power to carry out the reform. When this happens, great planners can implement profound plans. The greatest planner in modern times was Joseph Stalin. To accelerate industrial growth in his beloved Russia, Stalin set up a series of Five Year Plans that ultimately involved the movement of whole populations, wholesale reorganizations of economic systems and massive new projects on an unprecedented scale.

The ideal systems conceived by Stalin were hampered by two factors: the ignorant, selfish citizen, who did not always respond positively to economic measures brought about for his own good; and the physical environment, which did not always provide the required agricultural yields or other natural resources. To solve the problem of motivating people, Stalin promoted research already begun in social planning (best represented by Pavlov's success in getting dogs to salivate). To solve the problem of a not-always-co-operative environment, Stalin launched what turned out to be his last great project, the Plan for the Transformation of Nature.

Stalin, like Lenin before him, and Khrushchev and Brezhnev after him, thought in terms of big, centralized, uniform systems. But grandiose plans are not confined to the Russian brand of communism. Remarkably similar thinkers can be found in the free world as well. Edward Teller, the father of the American hydrogen bomb, is a proponent of geographic engineering (as opposed to civil engineering or geologic engineering), i.e. wholesale changes to the environment (through the use of peaceful nuclear bombs). This could mean the moving of mountains or the redrawing of coastlines to achieve economic goals. Meanwhile, W. Bennett Lewis, the father of Canada's CANDU nuclear power reactor, talks of using Nuclear Power Agro-Industrial Complexes (called Nuplexes for short) to turn the desolation of India's Ganges Valley into 100 million acres of rich productive land turning out food in torrents for the area's 200 million people.

Large projects are automatic consequences of large, centralized systems and large, centralized systems require a lot of advance planning. Unlike small systems, which can be built in many locations by many people in a relatively little time, large systems require a great deal of co-ordination and so a lot of administration, the centralization of great amounts of human and material resources and, consequently, a great deal of time.

Fifty developers in fifty cities can erect fifty apartment buildings of 100 apartments each (a total of 5,000 units) in three months. Ask one developer to put up one building of 5,000 units and the task becomes herculean, needing years in preparation and construction. (In Russia, where construction projects on this scale are not unknown, this is the very outcome.)

Were the North American building industry so slow it would lose its present ability to delay construction projects until a strong demand for them materializes. Developers would have to get out their crystal balls and predict what kind and size of buildings would be needed where, by whom, and at what price five years down the road. The industry—speculative enough as it is—would become downright chaotic. Where developers guessed wrong, and built homes that weren't needed, they would go bankrupt. Where they guessed wrong and didn't build homes that were needed, public outrage would be strong. In both instances government intervention would be demanded, either to protect developers or to protect the public. Regulators would authorize a systematic overbuilding of all categories of housing to ensure accommodation for future homeowners. Since this would be considered a public service, regulators would guarantee profits to protect developers unable to sell their houses.

Everyone would get his way—the only loser would be society through the taxes and bureaucracy necessary to manage mothballed houses in every community. The only reason for the loss would be the need to support a system which requires long-range planning.

As implausible as this scenario may seem, a similar development has already taken place in the energy industry, with large government-owned or government-regulated electrical utilities now expected to build 20 percent to 30 percent more generating capacity than needed on the hottest day of the summer or the coldest day of the winter—just in case one or more plants are shut down, or the planners' estimates are off and the extra electricity is needed. The cost of carrying that huge surplus is sometimes paid indirectly through taxes but more often directly through individual electricity bills.

Beginning in the 1970s, these costs started skyrocketing, largely because of nuclear systems. Nuclear systems require extremely long lead times, forcing interested utilities into building reactors long before their electricity would be needed. To the dismay of the energy planners, the future failed them.

The planners predicted in the 1960s that we would need twice as much electricity in 1980 as in 1970. The planners were wrong. For a utility like Ontario Hydro, more committed to nuclear power than any other in the free world, and so most vulnerable to human unpredictability, the result was a $12 billion debt and a doubling of the surplus—Ontario Hydro now keeps idle enough power plants to generate 50 percent more electricity than it can use. Worse still, Hydro's entrenched bureaucracy could easily repeat its mistakes of the past.

Ontario Hydro is now predicting that we will need twice as much electricity by the late 1990s as we did in 1980, and working toward the goal of supplying that need. Better planning methods may help dissuade them, but their previous record shows an unremitting resistance to change (all through the 1970s they could see that their predictions were off, but their system was too inflexible to respond).

In detail, planning methods seem highly complex. In concept, they are ludicrously simplistic. The forecasts of future electricity consumption are based on past increases in electricity use. Until the Seventies, electrical use doubled every decade with seemingly predictable regularity. In 1920 we used twice as much electricity as in 1910, in 1930 twice as much as in 1920 (and four times as much as in 1910), and so on. In 1970 we used twice as much as in 1960, and the trend was expected to continue into the foreseeable future and beyond. Based on this "historic growth rate," thousands of nuclear plants were planned. In the energy field, it was absolute heresy to think that the historic growth rate was fallible, aside from temporary aberrations such as war or depression. It had worked for the previous five or six decades and was expected to work for all time.

Imagine the disbelief and dismay in official circles, then, when a group of residents in downtown Toronto, after years of fighting, stopped Ontario Hydro from razing a block of homes in their neighborhood. The homes were to be sacrificed for a greater need—Ontario Hydro's planners had deemed a switching station necessary to meet the city's expanding electricity needs. That was 1971. The battleground shifted to a nearby location, at the edge of the residential area, and was later taken up by a city alderman who managed to convince other members of the city council that the switching station should not be built. City council successfully stalled Hydro's plans.

By 1977 the delays were making Ontario Hydro desperate. Insisting it had to build two transformers at the station immediately, a third for 1987 and a fourth for 1992, the utility flatly claimed "it is not possible" when asked if the city could do without them. Despite this, city council continued to deny permission to build the station.

By February 1978 those closest to the situation were at the point of exasperation. The mayor of Toronto called city council "irresponsible," adding that he was "badly shaken" that the council he presided over could so endanger the city. Toronto Hydro, meanwhile, was predicting the decision could result in blackouts in downtown Toronto that summer. The press was turning the heat on as well, with headlines ("Blackouts Seen") and editorials ("We Don't Want a Blackout in Toronto," "Toronto Can't Risk Power Blackout") denouncing the "blackout of common sense . . . an irresponsible and inexcusable act."

City council held firm. When summer came, the *Toronto Star* warned that power cuts would "soon be necessary and that more serious blackout risks exist in future." Toronto Hydro sent a warning to its customers, describing how and where their power would be cut (in a "series of 7, 15-minute rotating cuts") and suggested customers should keep one ear to the radio in case the utility needed to appeal to the public for a 25 percent voluntary reduction.

The letter, dated July 4, 1978, and addressed "Dear Customer," ended on a note that may have failed to reassure all customers: "We are advising you of this most unfortunate situation, not with the intent of alarming you, but rather to see that you may consider your own situation and make whatever emergency plans are appropriate to your own circumstances. We at Toronto Hydro hope that it will not be necessary to put this emergency plan into operation." And still city council would not budge.

There were no blackouts that summer. A little over one year later (October 22, 1979) the chairman of Ontario Hydro sheepishly sent a letter to the city acknowledging that after struggling with the project for ten years, and after "recent studies," they had decided to "defer indefinitely, plans for the . . . facility" because, with minor modifications to the existing system, they could "ensure reasonable reliability for the City's downtown lands until 1990 or thereabouts." The city councillors, it turned out, were right all along. The station had not been needed in the Seventies, would not be needed in the Eighties, and might only be needed in the 1990s, according to their latest planning forecast, which was also based on historic growth rates (but now with a slightly different history).

Though a surprise to Hydro, the city councillors (who after ten years had developed considerable expertise in the energy field) were not in the least astounded. When Hydro originally announced that electrical use would double over the next ten-year period, many wondered where the electricity would be consumed. Would the people leave twice as many lights on, watch two TV sets instead of one, run twice as many appliances as before? To meet government energy forecasts to the year 2025, in fact, will take some doing. First, the province of Ontario will need a population explosion, not of the modest kind once feared, but one which will fully double the population to about 16 million. Then, to have more homes to heat than the additional population would normally require, fewer people must live in each household, and three-quarters of all new households must be single-family, detached residences, which need most energy. None of the new homes will be better insulated, nor will other energy-conservation measures be adopted. Moreover, not one single homeowner will insulate an existing home.

Twice as many appliances will be used by each person, and none of them will be more efficient than the electric toothbrushes, ovens and refrigerators used today. Every person will also travel 25 percent more each and every day to work or to play, whether by car or by bus. The only exception will be airplane travel—everyone will have to fly twice as much as now. Of course, no car, bus, train, or plane will get better fuel mileage than at present.

People will also be twice as rich, so they can buy twice as many clothes, eat twice as much food, and own twice as many cars—all produced with the same inefficiency as today. But most new workers can be bureaucrats, employed in office buildings that leave just as many lights on as today, have no more insulation than today, and adopt not a single technological advance, whether or not the advance makes economic sense.

This is the stuff whereof official government forecasts are made.

The implausibility of that future led to the development of alternative planning methods, called "end-use planning," to show how much energy would really be necessary. This method, used by public-interest energy groups and virtually all critics of conventional energy policy, starts at some point in the future (instead of some point in the past) and predicts consumption for various energy uses at that time: how many homes will need to be heated, how many office buildings will need to be lit, how many industrial processes will need to be fueled. Then end-use planning shows how to most efficiently meet the requirements that will exist in future, with a precision previously unknown.

End-use planners lead us to this future using blueprints designed to produce that future. There is no question in the minds of most people promoting renewable forms of energy (or even most government and industry analysts) that conservation, solar power, wind power, and other forms of renewable energy have the potential to be superior in virtually every respect to non-renewable forms of energy. Similarly there was no question a decade ago in the minds of people supporting nuclear power (including present critics of nuclear power) that it was cheap, clean, and inexhaustible. Herein lies the danger. New information could come to light demanding radically different energy approaches. Should that happen, it will be crucial that society hasn't already locked itself into a course from which there is no turning back.

For this reason, end-use planning, so much more accurate than conventional planning, holds far greater dangers for society, especially when it is energy—so crucial to our lifestyles—that is being planned. To a great extent, planning for the future helps write the future. We built an extensive road network for our cars because we thought we would need it. And once we had it, we did need it, because society became structured around the car. Should we decide now that we don't need the roads, it would be too late for corrective measures, except at prohibitively high cost. The suburbs will not easily disappear, and the rich agricultural land lost to them cannot easily, if ever, be reclaimed.

We built an extensive, interconnected, electricity grid because we thought we would need it, and after we built it, we did need it, because society became structured around it. Had we our present knowledge, we might have opted for much smaller grids using locally generated electricity and satisified more energy needs with non-electrical sources. But once the electrical grid was built, and paid for, it became economical to use it for otherwise uneconomic uses. Quebec's enormous and still rising surplus of electricity has dictated electric heating for most new homes and forced below-cost sales of electricity to New York State. Like supermarkets caught with a surplus of tomatoes, our utilities are pricing electricity to encourage quick sale.

The cost is not only in dollars but in lifestyle. In the case of electricity, we now live with the threat of blackouts due to faults in the grid, pollution from coal-generated electricity and radiation from nuclear plants. Yet electricity makes up only about 15 percent of our total energy use, as opposed to the near total dependence the nuclear industry foresaw for us. Because its planning tools were so primitive (merely creating the electricity with the expectation we would keep consuming it) we had enough elbow room to pursue, for the most part, the lifestyles we chose.

But with end-use planning techniques, the possibility exists for a very close fit between planners' vision of the future and the realization of their plans. Had the nuclear industry used these techniques, instead of blind faith, it would have foreseen exactly how many buildings, lightbulbs, and TV sets would be needed to meet its own prediction, and then it could have facilitated their production. If sufficient new buildings were not being electrically heated, the industry could have offered incentives to change heating systems. If enough lightbulbs and TV sets were not being manufactured, it could have promoted new uses for them—lightbulbs for hydroponic gardening for example, and TV sets for beauty parlors and restaurants as well as bus stations and bars.

Whether practiced by the nuclear industry or the solar industry, the plans could have all been well-meaning, aiming to fulfill some pre-ordained, glorious destiny for humankind. In either case, we might have been forced to meet the needs of the system created for us—to conform to it instead of building systems that conform to our needs. Ontario Hydro has learned one lesson from its mistakes: in 1980, it formed an end-use planning department.

By its nature, planning, and particularly long-range planning, is anathema to a democratic system. No matter how democratically a decision is made, and how wide its consensus, if the decision permanently binds future generations, it stops being entirely democratic.

But long-range considerations do exist and cannot be ignored. They concern energy, the increasing age of our population, education, genetic engineering, changes in population, space travel, and others. The more we are asked to pre-package our future, the more we will be bound in a society where individual initiative and originality are seen as threats to the orderly advancement of a previously agreed upon future. The true challenge for planners should be to develop systems that eliminate the need for long-range planning—deplanning, in effect—so that, as much as possible, we can be flexible enough to decide on short notice whether a change in plan is called for.

Flexibility is generally related to size: the larger the project, the longer the lead time. Deplanning involves no more than recognizing the inferiority of large-scale systems as an often overriding factor and developing flexible or easily destructible systems in their stead.

In the case of electricity, the systems can be at least as small as car-engine-sized cogenerating units, of the kind now mass-produced by Fiat. If needed, millions of them could be manufactured and installed on short notice. Or the electricity systems could be solar powered, directly converting sunlight to electricity without the need of central electricity plants and the electrical grid. (These solar electricity systems are already cheaper in many applications, especially in those areas of North America without an already-constructed grid to distort the economics.) Many other small-scale electrical alternatives also exist which do the job at less cost, more reliability, and greater speed.

Small-scale systems also eliminate the need for large surpluses. Ontario Hydro needs a large backup for its present system to protect against the loss of the large plants in its system. Three or four separate breakdowns at a time, which can happen, could severely disrupt service. But if the components in the system are small, even tens of thousands of separate breakdowns would not jeopardize the system. For example, should 40,000 Fiat units break down the same day in 40,000 different buildings, the system would lose only as much power as produced by a single nuclear reactor.

Before Ontario Hydro turned to large-scale nuclear systems, the provincial surplus was 12 percent instead of the present 50 percent. Should Ontario turn to conservation and highly flexible small-scale systems, the surplus could safely drop to under 5 percent, saving tens of billions of dollars in surplus capacity and cutting electricity bills dramatically. Instead of the ten-year plans of the past, Hydro could have had one-year plans. Hydro could have waited until 1979 to decide how much electricity would have been needed in 1980. It could have avoided becoming a public nuisance, and saved all concerned a lot of unnecessary trouble and expense.

Putting off decisions until the last possible moment—because it allows for the accumulation of a maximum of information on which to base the decision—is basic to effective decision-making. Generals who stick to their original battleplans on the field go down in history for their blunders. Shopkeepers who prematurely select next year's merchandise invite bankruptcy. Quarterbacks who don't know enough to call audibles at the line of scrimmage soon play second string.

The principle of postponing decisions is instinctively understood by every individual. We avoid purchasing theater tickets a year in advance because of the difficulty of predicting our own interest, and our own availability. Similarly, though a wedding may be planned well in advance, the final decision to go through with the marriage is put off until the wedding day, to the relief of all those who've used the engagement period to spare themselves later grief. The more resolute lose nothing by using the time to reassure themselves of their choice.

Yet what is obvious to us as individuals often becomes obscured through training. We have learned to dread the convolutions associated with market-oriented economies instead of recognizing each small-scale, spontaneous convolution as a response to additional information fed elsewhere into the system just as the new convolution will itself become a source of information for others. In the process, we have destroyed a self-correcting mechanism, one which affects our health as well as wealth.

# PART 3

# After Shocks

# 18

## *The Energy Toll*

IF THE CANADIAN and American Cancer Societies were serious about cancer prevention, the first thing they would do is call for a comprehensive, crash program of energy conservation. There is no faster or surer way to cut cancer rates, nor is there any area in which more cancers can be prevented. According to medical statistics more people die of energy-related cancers each year than of tobacco-related cancers. Most of the annual increases in cancer may be attributable to increases in energy use, either as the fuel chemicals used to produce dyes and plastics or as actual energy production. Energy-related cancers may already account for more deaths than all other cancers combined. Had a crash program of energy conservation been started in the 1950s, about 50 percent of present cancers could have been eliminated. Were a crash program begun today, an even higher percentage of future cancers would be prevented, because rolled into the incredible number of coal- and oil-based cancers we have been facing are additional, possibly staggering numbers of uranium-based cancers being generated by the nuclear power industry.

Exxon is undoubtedly the world's number one cancer killer, followed by other energy giants like DuPont and General Motors. Yet it is only the tobacco companies that are repeatedly singled out for blame, out of all proportion to their level of responsibility. The closest that energy comes to being blamed for the historic increase in cancer rates is the general finger-pointing at "industrialization" as the culprit.

For this the credit must go to the extraordinary public relations skill and political lobbying power of the now integrated energy industry which has managed to cloud an otherwise clear association by steering research into the cures for cancer rather than into its causes. Cancer is not directly related to industrialization; only to excessive energy use in the industrial processes we've opted for. Industrialization without the proliferation of fuel chemicals and excess energy use has always been feasible, but the reduction of high cancer rates from these causes has not.

The failure to link increases in cancer with increases in energy is all the more remarkable when it is considered that the modern history of cancer is the history of energy production, dating back to 1775 when Percival Pott discovered that soot caused cancer of the scrotum in chimney sweeps. This was the first time that a carcinogen was identified and it was linked to energy use. Energy in another form—ultraviolet radiation from overexposure to the sun—was linked to skin cancer in the 1890s. In the early 1900s scientists familar with Pott's clinical observations extracted coal tars from soot and tried to induce cancer in animals. They tried the wrong animals. Luckier were two Japanese scientists who, in 1916, induced tumors in rabbit ears after dipping them daily in coal tar for over six months.

It took three more years before Western scientists would confirm the results of the coal-tar experiments, and even then they were viewed with skepticism. Cancer, according to the wisdom of the day, just "happened" when cells or tissues were irritated. The notion that one thing or another was more likely to make it "happen" took time to get accustomed to. The fact that carbon-based compounds like fossil fuels would find it easy to insinuate themselves into the cells of carbon-based animals like human beings has still not settled in.

In the 1930s, the Royal Cancer Hospital in London linked mouse tumors to tars extracted from pitch and oils and then proceeded to flabbergast the scientific community by producing tumors from a pure synthetic chemical that was a component of tar. Carcinogens, they found, were the products of incomplete combustion, i.e., of energy sources that did not burn cleanly. Meanwhile, in the United States, workers in the dye industry were commonly coming down with bladder cancer, and animal experiments confirmed that certain dye chemicals were responsible. The dye industry, then as now, was based on coal tars, and the carcinogenic chemicals were coal-derived.

A look at cancer maps of North America shows cancer rates to be highest in cities and industrial areas—precisely those areas which consume the most energy. Areas that contain petrochemical industries, like the U.S. Northeast or the Toronto and Sarnia regions in Canada, have higher cancer rates still. Three such districts in Los Angeles County, for example, have lung cancer rates in white males 40 percent above those in the remainder of the country, and it is estimated that pollution from petrochemical industries alone could account for 30 percent to 40 percent of cancers in the general population. In America, the states with the highest cancer death rates are all in the Northeast (with New Jersey the not surprising leader). The states with the lowest rates are the predominantly rural southern states (although with the movement of industry to the Sun Belt they are starting to catch up). The difference in cancer death rates between the five highest and five lowest states is over 40 percent. Cancer is now the number two killer, next to heart disease, on the continent and the only major killing disease on the increase. It is likely to become the number one killer before the turn of the century. It didn't have to end up this way. In 1900 pneumonia and influenza were the two leading causes of death. Deaths by cancer accounted for less than 4 percent of total deaths, ranked eighth behind "chronic nephritis" and "accidents." Out of every 100,000 people, cancer would claim sixty-four, compared to 175 today.

Many of those sixty-four deaths came from naturally occurring cancers: skin cancer from exposure to the sun; leukemias and other cancers from the background radiation in the environment; non-energy natural sources such as arsenic and asbestos. But many, perhaps most, came from the sooty environment of the steam age, especially in cities like Pittsburgh, whose coal-driven pistons polluted the skies with a pride unknown to later generations.

The leading killers of 1900—pneumonia and influenza—now claim only 3 percent of total deaths. Other leading killers of 1900, like tuberculosis and gastroenteritis, do not even show up in today's top ten. Medicine and hygiene have been able to contain these diseases or do away with them. The story of cancer has been just the opposite. New chemicals—many of them carcinogens—are being created at a rate of 25,000 compounds per year. Close to 2 million different chemicals have been made by us, and over 100,000 are in commercial use. Most have not been tested, nor would testing be feasible: to adequately screen one chemical for health effects alone at present requires two years of testing and costs an average of $250,000.

North American production of oil- and coal-based chemicals is approaching 200 billion pounds a year, and at present rates will double to 400 billion pounds per year in ten years. Annual production of vinyl chloride, the highly suspect oil-based carcinogen, amounts to about 7 1/2 billion pounds. Over one ton per year is produced of each of 9,000 different chemicals. Energy production in general—whether fossil fuels or electricity generation—has been doubling every ten to twenty years.

Cancer has not been contained because carcinogens have not been contained. Most diseases have a single cause—a specific virus or bacteria—that can be isolated, countered, limited, even eradicated altogether (as smallpox has now all but disappeared from the face of the earth). But for every carcinogen that is isolated, new ones are allowed to spring up; every cure found is made trivial as we realize that in the meantime we have fallen further behind; far from limiting production of carcinogens, industry receives huge government subsidies to produce even more.

While the chemical industry claims to be responsible for no more than 5 percent of all cancers, and circulates studies never published in scientific journals (through chambers of commerce and other institutions), it blames all lung cancers on smoking, all colon cancers on diet, and low cancer cure rates on foreign-trained doctors. The studies claim strong carcinogens are weak, and weak carcinogens have been controlled. And all industries but one nod approval. The maverick is the insurance industry, which must pay for the health costs of the cancer-producing industrial operations through the premiums it collects. For these purposes the insurance companies no longer rely on the petrochemical industry's data. Renewal rates on policies are sometimes raised by a factor of fifty.

The pervasiveness of petroleum products is awesome. Their carcinogens are spewed out of the exhausts of every motor vehicle, dispensed as drugs over the counter in every pharmacy, spread over the soil as fertilizer in every farm and garden, and then sprayed as pesticides over the fruits and vegetables we eat. They appear as food preservatives, food coloring, even as food itself in the form of milk substitutes and meat replacements. Petroleum-derived protein—15 million pounds of it per year—is now sold to us in baked foods, infant foods, meat products, and prepared foods. Of the 1,500 pounds of food we individually eat each year, nine pounds is nothing but chemicals added to foods—dyes designed for no other purpose than to spare us the shock of seeing the natural color of food (or the unnatural color of synthesized foodstuffs), preservatives to support an artificial system of large-scale food processing and distribution, flavoring agents to give some gusto to the petroleum mash that is easiest for business to process.

Reusable glass has been replaced by throwaway plastic containers, wood-derived rayon by oil-derived nylon, cellophane with plastic wraps, natural fibers with synthetic fibers. Each change from a solar to a fossil fuel was accompanied by imaginary or short-term benefits compared to the real and long-term harm brought down on us—but not revealed to us. Although the carcinogenic effects of vinyl chloride were discussed in detail at a 1971 meeting of the U.S. Manufacturing Chemists Association, the industry decided to prohibit publication as this would "lead to serious problems with regard to the vinyl chloride monomer and resin industry . . . and force an industrial upheaval via new laws or strict interpretation of pollution and occupational health laws." The need to suppress scientific data was stressed by Union Carbide's concern that in "areas most likely to be affected, such as food, food packaging, fiber and aerosols [Union Carbide] would be seriously hurt by arbitary or panic-induced government restrictions."

The following year, after additional research from several European chemical industries showed vinyl chloride to be an even more potent carcinogen than previously thought, the Manufacturing Chemists Association entered into a conspiracy with a European consortium to share their information but not to disclose it without prior consent—a pact that was put to the test almost immediately as the industry rejected a government request for information, deciding to honor its secret agreement instead.

As pressure grew for action against vinyl chloride and other petrochemicals, the industry responded with selected studies carefully designed or interpreted to produce the proper results. British Petroleum, in a 1975 study, was able to claim that the longer someone worked in a vinyl chloride plant, the lower the risk of contracting cancer. Exxon, Shell, Dow, DuPont, and others responded in kind, both refuting new information on the petrochemicals and presenting more studies of their own.

The industry subterfuge was exposed by a special committee report of the American Association for the Advancement of Science, the most prestigious scientific body in the U.S., which concluded that the Manufacturing Chemists Association "appears to have deliberately deceived NIOSH [The National Institute for Occupational Safety and Health] regarding the true facts . . . Because of the suppression of these data, tens of thousands of workers were exposed without warning . . . to toxic concentrations of vinyl chloride." The consequences of this conspiracy would show up as brain tumors in workers at the vinyl chloride plants of Union Carbide and Monsanto in Texas City, Texas, and as birth defects and liver cancers in workers and nearby residents of a vinyl chloride and polyvinyl chloride plant in Shawinigan, Quebec.

Had the full financial and health costs been recognized and accounted for, the fossil-fueled plastic revolution would never have occurred in the manner it did. Both workers and consumers would have rejected out of hand the health costs, higher prices, and no-win situations the industry by degree has enforced upon us. Internal contradictions, like delicious-looking-but-deadly food coloring, or flame-retardant-but-cancer-inducing sleepwear, would never have materialized.

Non-killing options are always available. Flame retardants for clothing became necessary as protection against oil-based synthetic clothing. Cotton and wool do not result in flash fires when lit. More expensive natural dyes, like red coloring from beet juice, would seem the bargain when the health costs of carcinogens like red dyes No.2 and No.40 were compiled.

The cancers that come from high energy consumption extend to many of those attributed to tobacco. About 20 percent of lung cancer deaths now occur among non-smokers, and the number is rising. The source is often energy-related, such as urban air pollution or radon gas (for people whose houses sit on top of radioactive landfills or who themselves live or work in uranium mining environments). Often it isn't one chemical that causes cancer, but the interaction of two or more chemicals. Without the intrusion of energy-related chemicals, many, perhaps most of the cancers attributed to tobacco would not have occurred.

The energy toll included far more than cancers. Black lung disease is still mentioned in the same breath as coal mining, but the sterilizations and mutations that arise from nuclear power are not as well known. Automobile pollution gives us not just cancer but various heart and lung diseases as well; benzene adds anemia to our lives as well as leukemia.

Acid rain is due not only to INCO's outpouring of pollution but also to emissions from coal-fired electricity plants and motor vehicles. Not only are our lakes dying now at the rate of seventy a week—fully 50,000 lakes in Ontario and New York State alone could become acidic over the next fifteen years—but agricultural soil is becoming eroded as well.

The economic costs of high energy consumption are staggering—government estimates for North America start in the tens of billions of dollars and go up from there. The costs of cancer alone are pegged at over $30 billion annually. Motor vehicle exhaust adds another $15 billion. No study has been done of the entire energy toll, including cancer and other health costs, the costs of imported oil, the costs of a dislocated economy, the costs of dead lakes and river systems, of impoverished soil, of denatured food, of pain and suffering and loss of amenities like sunlight, of planned obsolescence.

In comparison to all the net economic and human costs associated with energy consumption there are only net economic and human benefits associated with energy conservation (although it has been reported that an overzealous German insulated his home so well he died of asphyxiation). As it happens, the energy eliminated should an energy conservation program be undertaken would be those very energy forms most harmful to our health. Nuclear power—about 1.5 percent of our current energy consumption—can be terminated without many missing its electricity today or its cancers and leukemias twenty years from now. Coal and oil would be the next to go, as our society shifts to clean renewable forms of energy like methane gas and hydro-electricity. There are no additional cancers in trapping the sunlight that falls on our roofs, no additional cancers in using the waste heat from our industries and our power plants to heat our homes, and no cancers at all in the greatest energy source of all—the energy we save.

Use of energy products can be contained to allow medical science to make the same inroads on cancer that have been made on every other major disease. Technology, meanwhile, can begin to design systems with recycling in mind. No chemical can be indefinitely isolated from the environment—one way or another, it will escape from whatever containment system has been devised for it.

But products can be designed to be 100 percent recyclable, to form a closed loop so that virtually all chemicals used in an industrial process can be retrieved for reuse, so that all manufactured products can be recycled for their resources instead of sent to the dump.

This need not be an overwhelming task. When an alarmed U.S. government decided to reduce vinyl chloride emissions to one-five-hundredth of what had been permitted, the industry protested loudly. Consulting firms hired by industry predicted a financial burden as high as $90 billion and a social burden of over 2 million lost jobs. The Society of the Plastics Industry, in its plea before a Court of Appeals, stated, "The evidence clearly demonstrates that the standard is simply beyond the compliance of the industry . . . The general increase in raw materials costs [an unavoidable result of this standard] and the costs of monitoring, respiratory protection, medical surveillance, and record keeping all militate against the likelihood that the bulk of this segment of the industry would be able to survive economically. The evidence is clear [that] various industriesı would at the very least suffer economic disaster if not close down completely."

One year later the new standard—500 times more stringent than the industry had been used to—had been met without plant shutdowns or economic dislocations. B. F. Goodrich found that the initial capital cost of redesigning its manufacturing technology (which was only $34 million) was offset by its reduced labor costs and by leasing the new clean-up technology to other firms. Union Carbide expressed surprise at how easily the new standards could be reached, and pleasure at the unexpected benefits brought by recovering and recycling vinyl chloride, which would otherwise have been someone else's pollution.

In 1977, North American industry unnecessarily lost about a quarter-of-a-million pounds of vinyl chloride into the air alone. Now an industry leader in both the recycling and energy conservation fields, Union Carbide's new petrochemical complex at Sarnia, Ontario, is saving between 80 percent and 90 percent of conventional energy requirements.

Greater savings are still possible, and still necessary if industrialized society is to emerge untainted. Where industrial processes rely on carcinogens that cannot be completely recycled into useful or safe substances, the industrial process can be changed, as Atlantic Richfield and Goodyear found in switching away from benzene to safer solvents.

The most efficient systems-change of all, however, would be to the decentralized systems that are naturally energy conserving. A return to small-scale farming will eliminate the need for synthetic fertilizers and pesticides, to small-scale industry and urban economies most of the chemicals in common use, to durable products all the energy and chemicals used in making the same item twice.

Decentralized systems automatically solve most of the problems of pollution by not creating them in the first place.

# 19

## *The Energy We Eat*

THE SLEEK BRUNETTE bared all in revealing what set her apart from less sophisticated women. "I love to cook. But I don't always have a lot of time to spend in the kitchen," she asserted in a profound, full-page ad describing the typical *Cosmopolitan* reader.

> So I rely on convenience foods . . .
>
> I started off the day with powdered instant breakfast at least once during the past month. And 342,000 of the Cosmo readers did the same. We also bought 2,187,000 bottles of barbecue sauce, 750,000 hamburger or meat-mix packaged dinners, and 772,000 packages of rice mix . . .
>
> In fact, during the past month we purchased 1,808,000 frozen main courses and 1,054,000 frozen prepared vegetables . . .
>
> I am convenience conscious of course. But I'm also calorie conscious. That's why 668,000 Cosmo girls used low calorie liquid salad dressing in the past month. I also have a sweet tooth. That's why we bought 1,150,000 frozen pies and cakes in the past month . . . I guess you could say I'm THAT COSMOPOLITAN GIRL.

The scantily clad female of 1979, so liberated from her dependence on natural foods, may well be every technocrat's ideal. Memories of the day we would be sinking our knives and forks into a thick, rare cut of synthetic roast beef, prepared with just the right combination of chemicals and additives to assure proper texture, taste, color, and aroma, are so strong that many still suffer pangs of nostalgia, salivating at the thought. Others are still proceeding with the dream of eliminating the uncertainties in food with dependable substitutes.

"One primary target of fabricated foods is to create more animal and dairy type products without using animals and animal husbandry—to go directly from grain to a finished product that simulates what otherwise comes out of animal husbandry," a leading market research analyst for the food industry reported. "To meet this objective, processors are developing something that functions like a plastic cow with stainless steel innards and fiberglass teats. It feeds off petroleum and grain, and gives butter, whipped cream and even meat."

The plastic cow is being milked well. Margarine has captured two-thirds of what is now called the "table spread" market, and coffee creamers are no more likely to be fresh than fuel. Whipped cream is being whipped by corn syrup solids and hydrogenated fats, ice creams are being replaced by frozen lookalikes, and imitation cheese is taking over in school and hospital cafeterias. Many of our favorite chocolate products contain no chocolate, and the coffee bean may go the way of the cocoa bean. Thanks to the efforts of Procter & Gamble, and others, in developing coffee substitutes, we can look forward to the day of coffeeless coffee, with cellulose "grounds" giving body to artificial flavor, artificial aroma, and—to meet marketing requirements—laboratory caffeine.

To progressives in the food industry, this development would nevertheless be hopelessly old-fashioned, just more of the sad *status quo*.

"Until now, imitators have been content mostly to try to simulate the 'reality'—create imitation bacon that will crunch like, taste like, smell like real bacon, or imitation orange juice that looks, smells, and tastes like 'fresh squeezed' right down to the bits of pulp," an industry analyst said, lamenting the remaining links between food and agricultural products, and predicting instead, "completely new and fabricated products that don't even try to simulate anything already in nature. Gatorade and other glucose-based drinks have taken this leap, selling primarily on the basis of their own built-in benefits. Diet breakfast drinks and bars are also very well established examples of what we might expect in a laboratory-created, totally new form of food marketed on its own merits."

Taking the food out of food is not a consequence of science fiction aficionados running the food multinationals; it is merely a down-to-earth business requirement. The coffee bean has everything going for it except stability: a cold spell can wipe out the coffee crop in Brazil, overproduction in Africa can undercut prices at home. The price of pigs, corn, soybean, beef—every commodity that depends on climate—is similarly unstable and so unpalatable to large organizations that value security above all else, that have worked hard to develop a loyal clientele which might begin to look elsewhere for supplies. Corporations like neither to pass along unplanned price increases in times of shortage nor to explain why prices can't be cut in times of surplus. Their shareholders expect steady rates of return on their investments; divisions down the line need to be supported with assured supplies for their own profits; bureaucracies throughout expect a minimum of disruption.

If a $1 supermarket product has a 75-cent agricultural component, a doubling in price of that climate-controlled component raises the price to the consumer from $1 to $1.75. Consumers might switch to some substitute and when the price comes back down, they might not switch back. Previous years of advertising to create a market might have to be written off, and the public might have to be retaught benefits all over again. But if that $1 product contained only 10-cents worth of food, a doubling of that component would raise the price only to $1.10—a manageable increase which could even be absorbed by the corporation out of previous years' profits if necessary.

Nature has always been just too unpredictable for the food giants' liking and the trend to denaturing food has existed ever since the food industry became monopolized.

Like early consumers, the early food processors lived close to the food source and got their profits from collecting, processing, and marketing specific products from the farm. Most towns had a nearby mill to grind grain, and a butcher to repackage cattle as cuts of meat. At the start of World War I, North America had over 5,000 mills. A decade later the number was cut in half, with less than one-tenth of them handling over three-quarters of the milling output.

As corporate concentration increased and companies found they were no longer subject to the marketplace, they began creating their own markets. Margarine had been produced commercially since 1873 (having been previously concocted by a French food technologist who combined suet, chopped cow's udder, and a little warm milk to produce a better butter). A century earlier, beet sugar had emerged as the first major fabricated food when a German chemist discovered he could get sugar from carrots and beets. Now the development of fabricated foods would accelerate to produce artificial drinks and foods that required extensive processing, increasingly sophisticated packaging, more energy input, more labor input—anything to reduce the end product's dependence on nature. In the process, we learned the benefits of having cakes pre-mixed for us, and oranges pre-squeezed. When nutrients, taste, color, and texture were lost in the processing, they would be replenished with chemical substitutes to give us bread made bulky with wood fibers, and cereals laced with a battery of vitamins. Most of what we eat is now artificially flavored, and as a *Wall Street Journal* story pointed out, business has managed to do this for so long that we have literally forgotten what "fresh" tastes like.

Our memories will soon come back to us, and for that "COSMOPOLITAN GIRL" to remain fashionable she will have to boast about a completely different diet. Nature is returning with a vengeance. Since the disappearance of anchovy due to overfishing off the coast of Peru, nations have begun to value the natural food resources of the sea. Iceland traded volleys with Britain in their cod war; Canada and the U.S. have been embroiled in fishery disputes on both east and west coasts, Russian fishing vessels violating our waters are being arrested and Newfoundland is trying to fight off sister Atlantic provinces in keeping her catch to herself.

Concern over loss of agricultural land has heightened to the point where financial institutions like the Toronto Dominion Bank are warning of the consequences of continued urbanization. At present rates, even a country with the agricultural resources of Canada will have lost self-sufficiency in food within a decade.

The substitution of petroleum for natural produce—even should we accept the additional cancer load—will become a non-option as remaining energy supplies are diverted to purposes more essential than the marketing needs of the food processors. Substituting petroleum for human labor through chemical fertilizers, pesticides and fuel for overly mechanized farming techniques will also become unpalatable to a society aware of the energy constraints on our plates. The food production system, from the making of farm machinery to the flipping of pancakes, is totally dependent on energy—about 15 percent of all the energy we use goes into food. To maintain our foodstocks, we will need to tap that largest source of unused energy available to our food industry—energy conservation.

Convenience foods will become extremely inconvenient. On the whole, the production of prepared frozen foods like pizzas, cakes, pies, and TV dinners takes three times the energy needed to freeze plain, unpretentious fruits and vegetables. A simple fish-and-chips dinner, if home-prepared from fresh ingredients, saves about 4,000 kilocalories of energy per pound of food over the same dinner that came out of the frozen food section of the supermarket. For a TV fish-and-chips dinner, add 1,500 kilocalories for the energy that goes into making the aluminum tray, and another 500 for the energy that went into the fancy paper packaging.

Every time a family of four eats a TV fish-and-chips dinner it consumes the equivalent of about two-thirds of a gallon of gasoline more than that same meal would have required if cooked fresh—enough gas to drive a Honda about thirty miles. On a continent-wide scale, if everyone once a month forsook a frozen TV dinner for a fresh equivalent, the energy savings from this one measure would amount to about 12 1/2 million barrels of oil over a year. Forsaking only the aluminum tray would save about 3 million barrels of oil a year.

Another trend that will begin to reverse itself is our increasing use of sugar—now up to an average of 175 pounds per person per year, or a half-pound of sweeteners (mostly sucrose) for each of us every day. Some of our sugar consumption comes through natural sources, such as fruits and vegetables, but most of it comes from the marketing need to flavor the new generations of artificial foods we're being raised on. Packaged cereals, for example, commonly contain 50 percent sugar.

At the turn of the century we humans somehow managed diets that contained relatively little refined sugar, appreciating instead the taste of unmasked foods. Where sweeteners were called for, we ate fruit-sweetened cereals and pies. We now swallow ten to twenty times the sucrose our great-grandparents ingested.

Refined sugar gives us only "empty calories" that are devoid of all other nutrients while promoting tooth decay, contributing to obesity, and possibly contributing to diabetes. Our acceptance of refined-sugar diets—made possible only through intensive advertising—has drawn a bitter price in energy consumption. The processing of sugar beets and the refining of cane sugar (along with the milling of wet corn and the processing of malt beverages—also sources of empty calories) take 20 percent of all the energy used to process food on the continent. To produce a single pound of cane sugar requires 2,600 kilocalories, of beet sugar 4,400 kilocalories (ignoring the free energy of the sun in growing the crops). In exchange, we receive only 1,750 kilocalories back in the sugar's energy content.

Less energy is needed to ferment and distill sugar into alcohol than to refine it. Replacing just half the empty calories we receive from sucrose with less energy-intensive food would save the equivalent of about 15 million barrels of oil per year—and we would each still be consuming well over 100 pounds of sugars a year to satisfy our sweet tooths.

As our conventional fuel stocks become depleted, beef eaters may begin experiencing the kind of social ostracism that increasingly affects the cigarette smoker. Very few foods take as much energy to produce—the equivalent of about two gallons of gasoline is used for every pound of beef that is eaten. Measured by the nutritional value of beef, we need to invest ten gallons of gas to produce one pound of protein. Most of that energy is invested not in processing, refrigerating, or cooking the meat but in beef farms and feed lots. Our present methods of raising cattle by feeding them grains grown with petroleum products require ten times as much energy as getting the meat from the feedlot to the plate. Range-fed beef production is not much more efficient, but only because the modern cowboy has traded in his horse for the airplane and the jeep.

Chicken, on the other hand, is a far more efficient converter of grain, needing one-fifth as much energy as beef to produce an equivalent amount of animal protein. Deriving more of our protein through vegetables would be more efficient still. Although vegetables generally have less protein by weight than meat, more energy is needed to produce a pound of animal protein than a pound of protein from grains or vegetables.

Even bakery bread is rich in protein, one pound loaf containing fifty-four grams, or about the daily amount of protein recommended for adults. North Americans currently eat about twice as much protein as their bodies can use, most of it animal protein. If some of the protein we get from beef came instead from grain—for example by substituting over each month a pound of bread for a pound of beef (the equivalent of an extra slice of bread per day instead of an extra half ounce of beef)—North America would need to import about 130 million fewer barrels of oil a year from the Persian Gulf. If we also decided to substitute a pound of fish like perch for a pound of beef or shrimp per month, another 110 to 130 million barrels of oil a year would not need to be imported.

The more expensive the fish, the higher its energy content is likely to be since fishing vessels are willing to invest more time and fuel tracking down expensive varieties. Only six-and-a-half kilocalories of energy per gram of protein is used in catching herring, compared to eighty-one for tuna, ninety-three for halibut, ninety-five for flounder and 159 for king salmon. Harvesting and processing seafoods is almost as energy-intensive as raising and processing beef, with lobster, shrimp and blue crab needing virtually the same input of energy per output of protein.

Although small differences in selection of foods can result in huge differences in the energy we eat, the pressure to conserve energy will also lead to changes in what we do to food in the home. About 30 percent of all energy that goes into food (or 5 percent of all energy that goes into anything) is consumed in the kitchen—much of it inefficiently. The difference between an energy-conscious cook and a careless one can be as much as 50 percent, mostly in how thoughtfully the oven is used. Energy is often wasted in long warm-ups before the food is placed in the oven, in opening and closing oven doors (each opening can drop the oven's temperature by twenty-five degrees), and in leaving ovens on after the food is removed. Inefficient cooks also waste energy through poor use of refrigerators and stove elements, but because less energy is consumed there, the penalty is smaller. Baking for an hour in a conventional oven at 350° F uses 4,200 kilocalories of energy, eight times as much as simmering for the same amount of time on an eight-inch stove-top burner.

Kitchen equipment also counts. Dents in pot bottoms, improper lids, and the shape and material of utensils can make great differences. Pressure cookers cut cooking time—and energy consumption—for stews and pastas by about two-thirds.

Self-cleaning ovens (which are better insulated) require less energy to maintain a given temperature than a conventional oven, but those gains can be reduced or reversed depending upon how often the oven is asked to clean itself. Microwave ovens use about 40 percent less energy than conventional electric ovens but 30 percent more than top-of-the-stove cooking. Fridges and freezers can vary in energy efficiency by a factor of five, depending on how well they are insulated and how many time-saving-but-energy-consuming features (like built-in ice makers and water coolers) they have.

The dish chosen also plays a great part in determining how much energy we consume. Hamburgers need 50 percent more cooking calories than rump roasts; baked scallops use twice as much energy as sauteed scallops; baked salmon takes four times the energy of the same salmon poached; baked stuffed lobster requires six times the energy of boiled lobster. Everyone's diet has always been influenced by the price of the foods available. With energy constraints coming on, the consumer will have to weigh the energy costs as well. Canned salmon in a salad needs no additional energy: cooked on top of the stove salmon requires about 400 kilocalories per pound; baked in a soufflé, about 5,000 kilocalories per pound.

Many of these decisions would not have to be made if more efficient food systems were evolved, instead of our adapting to the inefficient food systems in place. Fridges and stoves could be made extraordinarily efficient if properly insulated and compartmentalized to allow the tomatoes and milk to keep cold when we reach for The prohibitive energy costs in packaging could be reduced if discarded packages were recycled into new products. Ninety-five percent of the energy that goes into aluminum trays and 74 percent of the energy in a steel can is saved when these products are recycled. Beef-eating becomes an environmental ideal when range-fed cattle using otherwise unproductive fields are tended without airplanes and fattened without chemicals.

We have striven to convince micro-organisms to convert starch into protein, and striven to perfect large-scale fermentation tanks for other food conversions. While these technologies have their place, the ruminant animal called "cow" is itself a small-scale fermentation tank with the additional advantage of being self-reproducing and mobile. Using energy, we have managed to increasingly centralize food production, thinking that greater efficiencies could be won with production done on a large scale. In the process of fighting the natural constraints discouraging centralized production, we have lost the natural efficiencies that come with decentralized systems.

Agricultural wastes make up the largest volume of North

America's organic waste materials. Several billion tons are produced each year, containing more protein than is available in currently utilized feedstuffs. But these wastes are usually spread too thin to make their collection worthwhile to centralized production systems. Too many different collection systems and processing systems—as a result of quirks caused by nature—would need to be developed. Small-scale systems, which do not have the inflexibility that comes with needing to feed large machines, can be better geared to using the resources in nature. And though each may be small, combined they have a large potential that is being blocked only by the centralized control of our business systems.

In Canada, for example, enough potato culls and potato wastes are produced each year—the equivalent of over 100,000 acres of grain—to supply sufficient high-energy feed to finish 200,000 cattle. One-quarter of these potato wastes are generated in the Maritimes, which is short on conventional cattle feeds but could be long on cattle were it not for the way Maritime business is handled.

In the Maritime provinces, food production, like much of everything else, has been taken out of local hands to be controlled by central Canada. In Halifax the butter comes from Kitchener, Ontario, the beef from the U.S. Many varieties of the fish daily caught off Maritime shores can be bought more easily in Montreal. The reason for these distasteful anomalies is always some artificially imposed system—either subsidized freight rates to help processors in Ontario undercut Maritime producers, or marketing practices that discourage small-scale entrepreneurs. The result is that vast, underused (or unused) local resources go to waste.

These losses are not unique to the Maritimes but occur in all regions of North America. Each case is different due to the different regional resources not used. Lands in the U.S. used for rice production—thanks to cheap diesel fuel that provides constant irrigation for otherwise unsuitable soil—may need to be turned over to uses less in conflict with nature. Thanks to that same increase in diesel fuel, which is raising the cost of shipping peanuts from Georgia to Canada, Ontario lands may become economic for newly developed peanut plantations.

Energy is a great equalizer. It can turn night into day and make deserts bloom. It can equalize differences in nature's systems, or differences in human systems. But when the energy used is non-renewable and the amount used exceeds our energy budget, the artificially imposed systems cannot be sustained. New balances, based on renewable energy budgets, inevitably must come about.

The phasing out of the agricultural component of food and replacing it with synthetics is another way of saying that the solar component of food has been phased out and replaced by non-renewable forms of energy. As non-renewable forms of energy become force-phased out of food, solar energy will again assume the dominant place it once held in food production and we'll become skilled at counting energy calories to keep food prices trim.

# 20

# *The Other Economy*

THE FASTEST GROWING sector of the market is made up of people who don't want to buy much. They can be seen shopping at flea markets and garage sales, admiring the clothes at second-hand boutiques, depositing money at credit unions and filling their own jars with peanut butter at the local food co-op or health food store.

They're learning to deliver their own babies, educate their own children, and defend themselves in court; they grow their own food, install their own insulation, put solar paneling in their homes, and read *Canadian Consumer* or *Consumer Reports* to find out which refrigerators and air conditioners best keep electricity bills down. When they're not bicycling to the subway station they're driving smaller cars that were purchased after inquiries into their mileage and durability. They often flatly reject overpackaged commodities, and prefer local products whenever available.

At supermarkets they dawdle over the generic foods section and make a point of purchasing their pop in refillable containers. The metal cans and glass jars that can't be avoided are often brought to recycling depots, and table scraps consigned to the compost pile. These people can be rich or poor, black or white, male or female, and they are now everywhere. They are all people with some degree of dedication to the conserver ethic.

According to a 1975 Harris Poll, when given the choice between changing their lifestyle to include less consumption of physical goods on the one hand and enduring the risk of continuing inflation and unemployment on the other, people overwhelmingly opted for a change in lifestyle. Over 90 percent were willing to eliminate annual model changes in automobiles and annual fashion changes in clothing. Wearing old clothes until they wear out (even if they shine) was OK for almost three-quarters; having a meatless day a week was acceptable to over 90 percent. Over 80 percent were willing to reduce the amount of advertising they were subjected to.

More significant than the number of people who might be prepared to take the pledge if presented with a clear choice is the number already thought to be practicing the conserver creed.

The Stanford Research Institute, one of the continent's most respected centers of business analysis, estimates that this sector of society had 15 million adherents in 1977, that it could quadruple to 60 million by 1987 and then could double again to 120 million by the year 2000. Of those 120 million adults, fully half would be strongly committed to the principles of conservation, and in addition to those 120 million, 25 million would be sympathetic to those principles, though not act on them. While this rapid and radical change in lifestyles could develop in fits and starts, it could also come upon us unobtrusively in the form of countless unannounced decisions made so inconspicuously that almost no one would perceive its progress until society had been largely transformed.

This latter course is precisely what has been happening. Although the oil crisis, gasoline lineups, international squabbles and nuclear protests have been attracting all the attention—comforting many by making the changes seem isolated—a large-scale, all-pervasive popular movement is underway, one having no headquarters, leadership, or co-ordinating body, yet affecting all social and economic areas, all geographic areas. Though the individuals, organizations and institutions involved in the movement are mostly unknown to each other, they are all moving in the same direction, acting apparently in concert. The direction is toward decentralization—toward smaller decision-making units and fewer authority figures.

This phenomenon, which has been attributed to "ecological awareness," "personal growth," and "material simplicity," is actually a great deal more profound and less pretentious than that. A self-help movement is underway. Rejecting large, centralized institutions, it is tied to no particular political ideology but intuitively recognizes that monopolies of any kind cannot respect personal preferences. The apparent return to "material simplicity" is a simple rejection of the inefficiencies of planned obsolescence. The return to "personal growth" is the necessary result of less dependence on impersonal institutions. The return to "ecological awareness" is the simple result of shedding external supports that obscured our connection to the environment.

But the movement that is underway is also a good deal less threatening ideologically than its righteous environmental trappings would suggest because it draws its support from the very many diverse interests which do not sympathize with environmentalists. It includes feminists, who've found no future in the conventional economy but have emerged as leaders in small business and the public interest sector where they can compete on an equal footing with males. It includes the gay community, which has stopped waiting for the rest of society to come around to its point of view and now helps itself by forming gay business associations and running gay candidates for public office. It includes (with equal significance) the millions of North Americans who now help themselves by pumping their own gas and jogging around the block.

People have found that leaving things to the conventional authorities will not keep gasoline prices low or their physical health high; that their personal satisfaction depends in large part upon having some degree of control over their personal lives. To assume control, they are depending less on the centralized institutions that evolved based on cheap energy and cheap resources.

Believing in economies of scale, and not needing to think about efficient use of resources, we built our schools, hospitals, and other institutions on a glorious and grand scale, requiring not-so-glorious grand bureaucracies to administer them. Having sunk so much capital into the buildings and the bureaucracies, we became so committed to making our investments work that we forgot their purpose. Innovative teachers and doctors became liabilities to systems that were set in their ways; students and patients became mere products processed through the production stream. The bureaucracy took on the role of "quality" control, setting standards to reject non-conformists.

North American bureaucracies have been less successful in stamping out human diversity than those involved in the Russian experiment with communism. Our centralized institutions were under strain even before the fuel crisis became evident. When the resource shortage hit in earnest, it became impossible to maintain these structures as before. The result of their failure is the rise of alternative institutions.

Lawyers have set up storefront offices to get closer to the community they want to serve and legal aid has become respectable for the most corporate of lawyers. Some doctors are busily practicing conventional medicine, but in community clinics, while others keep their traditional office but reject the traditions of crisis medicine, turning to acupuncture or specializing in new childbirth techniques for the home. Few professions have remained immune to this drive toward decentralization. Even policemen are returning to the beat after finding that their increasing rigidity was no way to deal with social change.

In countries like Japan and West Germany, whose citizens have a history of accepting centralized government, the education system is taking its toll. In West Germany each school year ends with over 500 suicides, and another 14,000 school-age youngsters attempt suicide. One out of every three people under the age of sixteen suffers serious symptoms of stress caused by school anxiety (for which the Germans have coined the word *Schulangst*) and one in five under sixteen is under psychiatric care. Ulcers are found in nine-year-olds; 20 percent of suicides under the age of eighteen fall in the ten to fifteen age category. According to one German educational authority, these figures should not come as any great surprise: "The education system is controlled by the bureaucracy, which is larger here than in other European countries and this system delivers exactly the candidates the bureaucrats want: ambitious, hard-working people who've learned the best way to succeed is always to agree."

North Americans are not as good at agreeing, and the result has been a massive rejection of the school system. Even in a heavily industrialized region such as the province of Ontario, students taking general-level courses geared to future industrial employment rebel. Throughout the 1970s, at least two-thirds of these high school students (general students make up 30 percent of the total) dropped out before graduating. Those that stayed in the system were often demoralized, and prospects also remained dim for those who graduated to college. University degrees have been so badly discounted that a bachelor's degree is often an embarassment, a master's degree considered a bare minimum and a Ph.D. an entry to a non-exclusive club of hundreds of thousands.

Any alternative to a sentence of school began to look attractive in the 1960s and 1970s, leading to the creation of literally thousands of alternative schools. But they were not embraced indiscriminately.

Like the automobile or many other innovations, most of the early "free schools" on the continent folded as the parents and students comprising the education market ruthlessly weeded out the ill-conceived, poorly-run, or unneeded learning institutions. The ones that remain are faring better than traditional institutions, many of them models of educational innovation and expertise. While the universities and community colleges in the Toronto area now compete for ever scarcer supplies of students (the University of Guelph resorting to a television ad campaign that packages itself with hype befitting a disco), learning institutions like the Skills Exchange of Toronto wonder how to accommodate their surplus of students. With 15,000 registrations in about 150 different courses offered over the course of a year, the Skills Exchange might seem to be akin to its university counterparts. But it operates out of a downtown second-floor walk-up with all of four rooms and a full-time staff of five.

Courses are chosen from a monthly catalogue distributed free in libraries, restaurants, and newsstands. The courses, which average fifteen students, can vary in length from one-day seminars to ten weeks of lectures; in content from the study of Shakespeare to bicycle repair; in size from two or three to an occasional fifty or more. Bureaucracy and capital investments are kept to a minimum. Its 150-odd instructors are responsible for providing their own classrooms and equipment; its students are responsible for evaluating the instructors. Most of the administrative work is carried out by mail, the staff of five ordinarily seeing neither instructors nor students.

The result is an intensely personal experience for students and teachers. Because the Skills Exchange can accommodate markets too small for universities to contemplate, options have been opened up for students with highly diverse interests and for teachers with equally diverse skills. The small classes are subsidized by the larger ones—no subsidies of any kind from governments or foundations. Yet the Skills Exchange attracts instructors of a caliber comparable to the heavily subsidized, also non-profit, conventional colleges. One attraction for teachers—a founding principle of the Skills Exchange—is free access to the skills of all other teachers.

The Skills Exchange has achieved its remarkable efficiency by substituting human labor for physical resources. Universities, to cut costs, buy communications equipment and pack 700 students into an auditorium to watch a video cassette of their professor; the Skills Exchange cuts costs by eliminating the communications equipment and the auditorium instead of the teacher. Universities need huge infrastructures to transport thousands of students in and out of some centralized location (including roads, special public transit routes, buildings, off- and on-campus services, administration); the Skills Exchange uses existing roads, public transit facilities, and classrooms during periods when they'd otherwise be underused. Universities need a computer to match teachers' schedules to pupils' schedules to availability of teaching space; the Skills Exchange simply lets teachers determine their own times and locations, and students pick from whichever are convenient for them.

In every case, a centralized system has been replaced with a decentralized one, saving money, physical resources and energy while providing greater flexibility and choice for all concerned.

Centralized learning institutions, like conventional universities, still have their place, but when there are no alternatives to them and they become the educational home for those who'd prefer to learn elsewhere, the system works to everyone's educational detriment. When the state decides to subsidize some learning institutions and not others, it works to the inequity of the education system in what amounts to arbitrary ideological distinctions and a violation of civil liberties. The French community of Penetang, Ontario, to preserve its culture, asked the provincial government to establish a French school for its children. The government balked, feeling it couldn't justify separate facilities for a relatively small school population even though members of the French community paid school taxes like everyone else. At the same time the province cavalierly pointed out that any group of parents can launch its own school as long as it can raise its own money. The community tried. In a highly charged atmosphere it set up school in an abandoned post office and airlifted teachers from across the province. The teachers donated their time, but the makeshift arrangements failed for lack of funds.

It was out of this kind of inequity and cultural arrogance that the idea of education vouchers sprang up in California; an idea long overdue that can redress the persistent political misuse of taxpayer money. Under the scheme it is the citizen who gets first claim over how his education money is spent—the government issues for every child vouchers for the amount that goes toward a child's education. The parents have the option of turning the vouchers over to the local school board, which would entitle the child to a conventional education in the public school system, or the parents can apply the vouchers to an innovative school. No one pays more taxes, but everyone gets more choice. Had this voucher system been in place in Ontario, the school children in Penetang would have been allowed the education their parents desired without the national scandal it created.

The principle of parent sovereignty over the education of children is nothing new. It has long been practiced in Denmark, a country widely envied for having one of the best education systems in the world. Any group of parents can start a school there, with the state paying 85 percent of the costs, the same subsidy that state schools receive.

It is no coincidence that people should be touchy about what information they're taught. As any stockbroker, bookie, doctor, pawnbroker, lawyer, or politician can testify, knowledge is power. It is the ability to move information—whether reliably or quickly or in new ways—that determines their degree of success. In a resource-short world, where the flow of information has been replacing energy use and staving off the shortages, control over information assumes new importance.

If there is one technology a conserver-oriented society depends on, it is electronics—the ultimate energy saver and the technology most intensively developed since the Second World War. Computers and other electronic systems do not just save training manpower in math tables; they save metal in every electronic machine, energy in every communications device. A million telephone calls can now be simultaneously transmitted through a single telephone wire and fiberoptics will allow the same amount of electricity to move many more messages while replenishing the world's store of copper with billions of tons of recycled telephone cable. New cars are being outfitted with microcomputers to tell us the mileage we're getting at any moment and new generations of typewriters allow a letter to be typed in one office and simultaneously reproduced in another. Editions of newspapers are now being shipped by communications satellite in pre-printed form to save on transportation costs, giving Chicago readers of the *New York Times* or Calgary and Ottawa readers of *The Globe and Mail* their daily newspapers by seven o'clock in the morning. "We're simply substituting a satellite for airlines and the Post Office," explained the *Globe*'s publisher.

Computers, which originally needed rooms of their own, can now be outperformed many times by microcomputers on a silicon chip the size of a fingernail, allowing information to be processed anywhere, allowing information to be disseminated everywhere. The result has been a massive decentralization of even the biggest of businesses, which find that decisions can be made locally with better results and at lower cost by personnel who have access to the same information as head office.

Westinghouse, Dow, General Motors, General Electric, and others are decentralizing with the aid of electronics, and enterprises that use electronics to become further centralized, such as Canada's highly automated Post Office—are collapsing, bogged down by their own weight and losing customers to their decentralized competition which can deliver letters at half the price and twice the speed.

Telephone equipment, unchanged in eighty years, became a remarkable new instrument of communications after the famous ten-year Carterfone battle against AT&T and Bell decentralized telephone company control to allow the interconnection of private devices to the telephone network. A new communications industry was created overnight—automatic call forwarding, memory phones, automatic answering services, new styles in telephone handsets and literally thousands of other innovations were conceived, designed, patented, produced, and marketed as soon as entrepreneurs were given access to the market.

More revolutionary business breakthroughs will occur in entirely new ways of doing business. A highly decentralized form which emerged in Toronto in 1980 may be one successful model for doing business in future. Called Trade-A-Skill, everyone from accountants and computer programmers to gardeners and interior designers are using this remarkably flexible system to market their business skills. But instead of marketing their skills for money, Trade-A-Skill members directly obtain the goods and services their money would buy. For example, a plumber would be credited by Trade-A-Skill for every hour spent unclogging the toilets of members referred to him. The plumber, meanwhile, could call on the organization to provide him with the names of a graphic artist (if his business needed new letterheads designed), an investment analyst (if his child needed tutoring).

The plumber makes his own arrangements with the people he contacts; for every hour of work he receives, Trade-A-Skill debits an hour of credits he had accumulated. If the plumber needs one of the non-professional skills offered, such as furniture stripping or thoroughbred handicapping, only half an hour of his time would be debited.

The services, published through a directory of skills, provide a non-intrusive, non-bureaucratic and highly efficient method of operation.

Trade-A-Skill acts as an information middleman that eliminates many of the resource intermediaries otherwise required. Whether on the giving or receiving end of the services, members have the flexibility of participating on any schedule and in any fashion they choose.

For some, Trade-A-Skill provides a supplementary income to a nine to five job: for others, it allows them the individual freedom to work odd hours, or in fields for which conventional employment has becomes hard to find. Unemployed welders and prematurely retired executives have additional options to utilize their skills in exchange for the works of others.

The idea of trading skills was adapted from the Skills Exchange, new work was fashioned from old, decentralization made possible through different ways of organizing information.

The decentralization is so complete that cottage industries are once again becoming competitive after centuries of increasing obscurity. One-third to one-half of the workforce in some advanced manufacturing centers could already perform their work at home. And in Europe, where gas can cost sixty dollars a fill-up, companies like Wolfcraft (which makes power tools for the do-it-yourself market) have reinvented the cottage industry but with a high-technology base. The company saves factory space and administrative costs and gains flexibility; the worker saves gas and traveling time while gaining personal freedom. The trend to job sharing, which has been accelerating only in professions like teaching, where husband-and-wife teams sign on to share one teaching position, can now be extended to the home, with work teams formed in families or among friends, in an atmosphere alien to the still pervasive punchclock production line.

The family as the basic unit of society has been structurally weakened since the advent of fossil fuels robbed the solar-fueled family of responsibility for production. Many family and neighborhood functions such as education or care for the young, the old, the infirm and the unemployed, were displaced by larger institutions and governments. These shifts were the social aftermath of a changing energy and technological structure designed to tap the benefits of greater energy use, a substitution of apparent wealth for real social stability. This substitution is still promoted. Government leaders tell us to damn the health costs if necessary to expand energy production; blinkered economists tell us divorce is good for the economy, and cite the construction of 2,000 extra homes in Calgary as a positive result of the shambles of 25,000 marriages. Even resource-rich Alberta, whose Heritage Fund rises in direct proportion to the depletion of its mineral resources, must soon return to the solar family exploring for energy efficiencies and economic liberties.

Freedom from regulation has also come to the broadcast media, where the proliferation of television stations will profoundly change the nature of modern culture. As happened in the magazine market, where the mass circulation *Life, Look,* and *Saturday Evening Post* went out of business to be replaced by tens of thousands of highly specialized publications, the public palate will reject the current pulp designed for a standardized audience of thirteen years of age and up. TV markets once considered unthinkably small (as highly specialized to the TV networks as Skills Exchange courses are to universities) are already claiming their share of air space. Through satellite networks, North American audiences can already tune in to TV channels exclusively dedicated to childrens' shows, or to sports events, or to religious shows, and future channels will be geared to opera buffs, foreign language markets, and science fiction aficionados.

In areas like Toronto, approximately a hundred stations are already available by satellite antenna, to be followed by hundreds more, changing our definition of what comprises a large television audience. With 1 or 2 percent of the market a profitable share, a cellist, a Shakespearean actor, a ballet dancer or a poet can each find work on a regular basis, encouraging more cultural diversity and giving less economic power to any single entity. General Motors would need to place hundreds of ads to reach the precise audiences currently reached imprecisely with one, and it may not seem in its interest to sell Cadillacs on a cleric's station, sub-compacts on a Fortune 500 station, gas guzzlers on an environmental station or any car on a children's station. Inventors with a new product of interest only to home handymen, or butterfly collectors, or political pundits will now have a relatively inexpensive means of reaching the precise market they're looking for. Paralleling the recent history of magazine advertising, the cost of reaching each viewer will skyrocket as markets become more and more fragmented, but because viewers not likely to want a butterfly net or crystal ball aren't reached, the cost of reaching the wanted viewers will plummet.

So powerful are these electronic information technologies that major efforts are underway to monopolize them. IBM, IT&T, Bell, CN-CP, CBC and CTV, the cable companies, and others are all involved in suits and countersuits, jockeying for position. To the victor of the communications battle, as to the victor of the energy battle, will go much of the spoils of the next century, unless the monopolies can be curbed by the very movement toward decentralization spurring the new alignment of economic forces.

Opposing the mammoth monopolies will be a fundamental new organizational form that has emerged, a force organized around the nearly one million public interest groups active in North America. Environmental groups, consumer groups, church groups, civil rights groups, and women's groups represent a sophisticated revival of grassroots democracy. Their power—though still hard to perceive—is becoming too great to ignore.

# 21
# *Thermodynamic Government*

JOE CLARK GOOFED over the eighteen-cent election.

In retrospect, everyone agrees it was foolish of his Conservative Party to risk an election over a budget whose best news was an immediate eighteen-cent gas tax, followed by regular fuel hikes after that. Who, after all, is crazy enough to vote for higher taxes? But, believing that the tax was indispensable for the country, the Conservatives, in the words of Prime Minister Clark, "decided to trust the Canadian people with the hard decision" that had to be made.

To bolster their case, they went to the business community and got its backing. They went to leading newspapers and got their backing. They went to economists and got their backing. The Conservatives went and got the support of almost everybody except the public. They lost the election, and Pierre Elliott Trudeau's Liberals were swept back into power with a majority government on their pledge not to raise gas prices dramatically. Of course, as soon as the Liberals were elected they proposed almost identical increases. The Liberals, too, secretly backed Joe Clark's high gas prices.

Anticipating their defeat before election day, the Conservatives warned that it would be a black day for democracy if voters couldn't be trusted to take bitter medicine when it was called for. The implication was that governments lose the ability to lead and pursue short-term policies to get re-elected to the country's long-term detriment, eventually threatening democracy itself. Yet the Conservatives had so little confidence in their ability to sell their budget directly to the public that they didn't even try.

After the election, the Conservatives fully recognized the danger of trying to forthrightly lead an uneducated public in a democracy. It cannot be done in hard times, except at great political peril to anyone proposing unpopular measures. The record of most democratic regimes around the world is a short and unstable one. The tendency of leaders in a democracy is to become increasingly cynical and authoritarian, to treat the public as an obstacle to good government and to rely on simplistic arguments or the voters' poor memory when passing unpopular measures. As the Conservatives implied, the long-term stability of democracy is indeed in question. And the alternatives become either to live with bad government until we are forced into some other form of government, or to create an informed citizenry. Education becomes an absolute requisite to democracy.

To Clark's misfortune, there was no public lobby advocating higher oil prices. Only the oil producers (not highly credible with voters) were actively championing higher prices.

Yet Joe Clark did not have to lose the election. Had he considered higher gas prices (and his government) important enough to launch a national debate over the issue, involving thousands of different communities across Canada in thousands of different ways, his Conservatives would still be in power. A majority of the public might not have been ready to agree with him but enough doubts about the wisdom of subsidized prices would have surfaced that the opposition parties wouldn't have dared topple him.

The issue is not too murky for the public to fathom. Several years ago in Sweden a far more complex issue—nuclear power—was debated by the public. With a minimum of government assistance, 9,000 groups became passionately involved in the subject, educating themselves and forcing politicians to speak in greater specifics about a subject people were now equipped to evaluate. In 1980 the Swedes held a referendum on nuclear power—only the fourth referendum in their history—with all observers remarking how civilly both pro- and anti-nuclear sides had carried out the long campaign.

The North American citizen is also no novice.

Although governments and industry here have done their best to keep information to themselves and people in the dark (all the while deploring public apathy), the public has demonstrated a remarkable ability to deal with the most sophisticated of issues. No government in North America would want to compare its record to that of the public interest groups—not in consumer protection, not in civil liberties, not in safety standards. Very little legislation that governments are proud of was proposed or passed without a public lobby first demanding action—in the energy field, for example, government action perfectly parallels public demands of several years earlier.

Public interest groups represent the profoundest educational instrument in North America today—their effectiveness dwarfs the efforts of all the high schools, community colleges and universities combined. Starting with a population wholly supportive of nuclear power, the public interest groups—in less than ten years—had informed and reversed the opinions of half the population of North America. This conversion of perhaps 125 million people—accomplished with almost no money and against the weight of all governments, the bulk of media, and virtually all corporations and labor unions—can be primarily attributed to no other source. The press became interested in the nuclear debate in earnest only after Three Mile Island, when, except for the very few journalists who had become knowledgable in the nuclear issue, most of the media needed to turn to the public-interest sector for information. Worse accidents than Harrisburg's had occurred before the anti-nuclear groups' message was publicized, but they went unreported by the press.

Anyone visiting the USSR understands what apathy is, recognizes that the North American spirit has never been successfully subjugated by force or by training. But when events seem beyond our control, our involvement declines accordingly.

Ask someone to join you in stopping an all-out nuclear war and he may grudgingly send off a letter of protest to a political representative. Ask him to help stop a nuclear reactor that's being built on the other side of the country and he might send a cheque to an anti-nuclear group as well. But tell him a nuclear waste disposal dump is being planned for his community and his complacency will probably disappear. The difference is not so much that it affects him personally, because nuclear war affects him personally too, but more that he's in a position to act: strategically, he knows the lay of the land, who is likely to support him, who can be persuaded to his point of view, how to reach local service clubs and church audiences, which local politicians should be lobbied, whether placards and demonstrations will be productive or counter-productive, or whether neighboring communities will be supportive—the problem that he's been presented with is on a small enough scale that he can comprehend its bounds and deal with it, as he may have participated in choosing the location of the town's community recreational center, or deciding whether a certain traffic intersection was safe.

The further from us decisions are made, the less likely we are to become involved. Each tier of government becomes, effectively, another barrier between the citizen and his ability to have control over his own life. The more that powers are handed down the government hierarchy—from federal to state or provincial, from state or provincial to municipal—the more democratic our society will become. Unnecessary levels of government, such as the regional municipalities or metropolitan governments introduced in the last few decades, should be dissolved.

There is no reason for most of our government bureaucracies, many of which overlap and all of which now represent a fourth force in government, unanticipated by our constitutions. The checks and balances built into our political systems with the executive, legislative and judicial branches have all been thrown off kilter by burgeoning government bureaucracies which have managed to make our judges, senators, members of parliament and presidents all work to their pace.

The bureaucracies represent little productive work and much agony to all but the bureaucrats. It is they who have assumed the role of regulators—a role almost independent of the political process. Politicians make the laws but then it is left up to the bureaucrats to interpret them. A "clean" environment, a "safe" automobile, or "adequate" health care can mean whatever the bureaucrats decide, giving them license to set standards and enforce procedures in almost every aspect of life. The bureaucracy has succeeded in alienating both industry (through often unreasonable regulations) and the public (by not enforcing the regulations industries will not meet).

Compounding the problem in specialized industries is a built-in paradox—those most competent to oversee the industry tend to be in the industry itself. Government recruits as regulators people who owe their careers to an industry that they may be returning to. In cases of extreme specialization, such as in Canada's nuclear industry, the situation has deteriorated to "a cozy network of old boy relationships," according to a study by the prestigious Organisation for Economic Cooperation and Development, which represents the Western industrialized nations.

The public now needs protection against the government; even the regulatory agencies presumably protecting the public interest become openly hostile to their constituents. The former president of the Atomic Energy Control Board, when asked at a royal commission hearing why the public wasn't apprised of safety matters that could affect their health, replied, "because I don't think . . . it is any of their business." His successor, in testimony before a Select Committee of the Ontario legislature, branded all nuclear critics alarmists and obstructionists.

Government regulatory bodies have so impeded the functions they were created for, and so closely are their interests identified with those they're intended to watch over, that the regulatory bodies themselves have become the objects of sit-ins, demonstrations and other public protests. When the regulated become the regulators, it is time to question the very integrity of the regulatory function and to demand a scheme less susceptible to corruption. A mechanism already exists which offers a greater level of protection. It is self-correcting and doesn't require people to watch the people who watch the people who are being watched.

Rather than this army of regulators, industry could be asked to regulate itself through the insurance companies, whose business it is to determine risk. The passage of only one law, which would require all industrial products and processes to be fully insured against any possible present or future damage to property or person, would satisfy the public a good deal more than the good intentions of government—without need for any regulation. The insurance companies can be counted on to look after themselves.

To keep its premiums low, a chemical company would now have to convince an insurance company that a new compound was safe. The riskier the compound looked, or the more questionable the information was, the higher the premiums would be. When a product represented a risk so great that no insurance company would insure against its possible consequences, as is the case with nuclear reactors, the product would be deemed too great a risk for society to assume. It could not be marketed until the risks came down to at least the point a speculator like Lloyd's of London dared gamble on it.

In so doing, the manufacturer would only be paying his own way, only making sure others aren't later to suffer financially from decisions the manufacturer made. The advantages for society would be immense, and not only because the bureaucracy could be dispensed with. Right now, because industry does not pay for the amount of regulation it necessitates—or the amount of damage it creates—there is little incentive to think of systems that save others money. But make those costs the company's own and corporate entities will work toward eliminating those inefficiencies the way inefficiencies elsewhere have been eliminated. A revolution in innovation will occur, leading to safer, stronger, more durable products whose benefits will show up in society's ledgers. INCO saved itself some research funds decades ago by installing mining systems whose pollution cost the Sudbury district $500 million per year, but cost INCO nothing. In 1980 the company announced that by chance, a worker had an idea that not only will significantly cut pollution in future but will save it money in the process. Forced to clean up after itself or pay compensation, INCO would have long ago done something to avoid the $500 million annual expense. It would have made this discovery and others on its own, needing neither luck nor environmentalists nor government regulators to help it.

Achievements by other firms in other fields would be as startling.

Gone would be the regulatory bureaucracies involved in the environment, consumer protection, urban planning, and other departments. The small groups of advisors that presently aid politicians in policy matters would be all that remained. Governments would still hire researchers and consultants in universities and industry, but without a mass of bureaucrats prepared to resist any change that politicians decide to pursue.

The government bureaucracies involved in social services can also be replaced with far greater efficiency and far lower cost. Voluntary organizations, whether church-run or community-based, are no less compassionate than the government agencies presently looking after senior citizens, runaway youths, the handicapped and the unemployed. Government social services are recent phenomena that have only added alienation to the other miseries of the disadvantaged.

In cases where people are unable to help themselves—such as those who need the care of institutions or children who need foster parents—voucher systems similar to the education vouchers in

California could be used to guarantee that sufficient funds are allocated by taxpayers to each area of need. Where taxpayers do not favor private social agencies the vouchers could revert to the traditional government bureaucracy.

But in most cases, the entire government bureaucracy can be eliminated by bringing in a guaranteed annual income (or negative income tax) system to give people directly through the income tax system the money they've been getting indirectly from dozens of different agencies. The welfare function (except for emergency relief) could now be disbanded. People themselves would become responsible for how they spend their funds. Unpredictable costs, such as medical expenses, would be controlled through medical insurance plans—payment of the premiums would be considered a social right. We should no longer be in the business of dictating what percentage of public welfare funds goes to food or what percentage into accommodation. This paternalism has limited the ability of people to innovate with the scarce resources they have while infringing on civil liberty and personal dignity.

A negative-income-tax system can be highly complex or simplicity itself; it can be so straightforward that every taxpayer would regain control over his income tax forms, eliminating the need for accountants and other newly necessary middlemen like the H & R Blocks of this world. For a negative income tax to work, society would have to set an income level below which no member of society should have to live—perhaps the poverty line, which is already measured by various agencies, and varies from region to region, depending on local factors. A family of four in a city might need $10,000 per year to live without undue hardship.

Those earning over $10,000 would be taxed only on the amount over the $10,000, preferably at a flat rate like 50 percent with no exemptions. A family earning $20,000 would pay tax on $10,000, or $5,000 in all (an effective tax rate of 25 percent on the family's entire income). A family earning $30,000 would pay 50 percent on $20,000 of it, or $10,000 (an effective tax rate of 33 1/3 percent). A family earning $50,000 would pay 50 percent of $40,000, or $20,000 (a rate of 40 percent). Without all the complications of the present system, but with great equity, this taxation system would bring in the same amount of revenue from the same income classes as our present convoluted system (although large loophole users would be big losers with the elimination of all exemptions). As government bureaucracies are reduced and government expenses decline, the 50 percent tax rate can be lowered.

Those living below the poverty line would still fill out tax forms, but instead of paying money to the tax department they would receive money back. A family with no income would receive $10,000; one with $6,000 income would get $4,000 back. Instead of picking up monthly cheques from the welfare department, the cheques could be mailed from the tax department. Overpayments and underpayments could be easily cleared up on the next year's income tax forms. As the poverty line changes due to inflation, so would the amount of taxes paid out by taxpayers or paid negatively by government.

The negative income tax allows for the abolition of the minimum wage, which would encourage companies to compete for employees on the basis of job quality, work environment, and job experience. Jobs that might not otherwise have been available could now be offered. An individual might well opt for a lower-paying position or an apprenticeship position that offers excellent future prospects in lieu of the higher paying, dead-end jobs so common today. There would be no danger of employer abuse because individuals would have no reason to accept low-paying jobs unless other work benefits provided compensation. There would be cases of individuals who are content to do no work and live off the state (as is the case today) but these would be few and become fewer. Studies in Canada and the U.S.A., where whole communities have lived on a guaranteed income basis for years, show no deterioration of the work ethic. Although people could sit back and collect a modest amount of money, even those who would benefit financially from that arrangement chose to be productive members of society. A negative income tax scheme—by giving people more options—can only lead to a strengthening of the work ethic, especially since it will most strengthen the entrepreneurial sector. Home industries—everything from sculpting to bookkeeping to furniture making—can be undertaken with the security that a year or two of below-poverty-line living will be supplemented. Speculative businesses and services can be dared. Individual inventors—who still come up with most of the technological advances adopted by industry while industry research departments continue to get all the government subsidies—will be able to devote their full attention to their main preoccupation, instead of having to support this indispensable work with less productive and, to them, meaningless work.

Although the income tax system would allow no exemptions, legitimate business expenses would, of course, need to be deducted. For corporations, these would include rent, raw materials, advertising and promotion and other costs directly associated with running a business. They would not include political contributions—politics should be the business of people and the public should be encouraged to take its business seriously. Individual donations to political parties should be considered legitimate tax-deductible expenses made in the public interest, as should all other expenses made in the public interest.

Social entrepreneurs—those starting new public interest groups to meet new social needs—will now have the same economic status as business entrepreneurs. More people will be encouraged to enter the public interest sector, although, like the entire negative-income-taxed society, the marketplace will determine whether new groups survive. Most don't, but the ones that do are responsible for some of the lowest priced, highest caliber research available in areas as diverse as consumer protection, arctic wildlife, energy policy, the regulation of nuclear reactors, the treatment of political prisoners abroad, civil liberties at home, ratemaking of telephone companies. This research is now published by bodies like the Economic Council of Canada, presented before Royal Commissions and Senate Committees, contracted out by government departments, relied upon by the political opposition, pounced upon by the information-hungry press.

The principles of thermodynamics give us insight into how the universe operates, and tell us that efficiency is a measure of how well the work that needs to be done is matched to whatever is doing the work. A family moving to the West Coast could transport themselves via Allied Van Lines and have the movers bring their belongings in the family Honda; eventually, all the work that had to get accomplished would get accomplished, although the match would not be very efficient. A house could be heated electrically, which requires that energy be brought to an electricity generating system from a coal or uranium mine, that it then be used to generate heat of thousands of degrees producing electricity that's transported at exceedingly high voltages over exceedingly long distances to be stepped down to lower voltages for use in the home where it is then degraded to 68° F heat. Or the home can be heated by the rays of the sun, which arrive on the spot ready for use at temperatures close to those required. The electricity we would save could be used for all the high-level needs solar heat can't meet, such as communications and lighting.

The same laws of thermodynamics could be applied to government. The Federal Government in Ottawa now tries to help an undeveloped region in the Maritimes get jobs by giving a grant (using money from Alberta) to a company in California so that it will relocate on the Atlantic Ocean and export to the United States. Instead the money could go directly to the region that needs it, to help it find its own solutions. Instead of three levels of government being involved in health matters, and trade matters, and environmental matters, and energy matters, more and more of these decisions can be left not only to lower levels of government but directly to the public. The public is collectively far less volatile, far better able to discriminate, far more reasoned and receptive, and far, far, wiser than the governments it gets to choose between.

Governments have been willing to let themselves be stampeded by special interests or persuasive individuals into countless acts contrary to the public interest. They have ordered mass innoculations for swine flu, authorized all-out nuclear programs, built roads willy nilly, failed to question the widespread use of DDT, permitted the mass spraying for spruce budworm, supported (then withdrawn wholehearted support for) new math for schools, centralized all of society, created the suburbs, and embarked on the Vietnam War. These examples of great inefficiency often involved the deception of a usually skeptical public—the public has repeatedly shown itself to be a force of moderation, rarely making mistakes on its own. Though sometimes fooled by misplaced trust in politicians, the public is quick to rebound when the deception is uncovered. A political party can casually choose a Goldwater or a Reagan as a presidential nominee but only under extraordinary conditions could someone at a political extreme be chosen as a national leader by the whole of the people. When the public is asked to decide by referendum on matters too touchy for politicians to handle, it responds with rational decisions unbiased by allegiances to political friends and unclouded by political dogma and ideological categorization. Austria, in a 1979 referendum, decided by the slimmest of margins to reject nuclear power. Nearby Switzerland that same year voted in a referendum to proceed with nuclear power. On a less emotional issue, Calgary and Edmonton were both faced with municipal plebiscites over the construction of large civic projects. Though the cities are located a relatively short distance apart, and the plebiscites were held within a short space of time, differences in the two municipal mentalities voted one project in and one project out.

By the same token, having no party line to follow and no vested interest in the bureaucracy permits innovations to get wide public discussion and sincere consideration. Even Proposition 13, which was billed as a black mark on citizen referenda, had none of the negative effects predicted by political pundits, who failed to appreciate the public's awareness that what can be voted in can be voted out, that no irreversible damage could be caused by a change in taxation targeted at bureaucrats but that real damage was being done by the *status quo* imposed by bureaucrats. When the promoter of Proposition 13 tried in a 1980 referendum to slice taxes even further, the public—as if to punctuate its reasonableness—politely declined.

Quebec's referendum on its status within Canada received the fullest and fairest hearing, and led to much better understanding of the issues dividing Quebec from the rest of the country. The education process that the people of Quebec underwent—three weeks of prime-time television debates from the legislative assembly building, followed by a two-week cooling off period, followed by a referendum campaign thirty-five days long—dominated conversations and electrified the population. It epitomized the very best in democracy. Far from having a deleterious effect, airing issues of such importance can ultimately only strengthen Canadian Confederation by coming to grips with the need for the country to decentralize, to let more of the various regions have more of a say in their own destinies. Quebec is not alone in its frustration with centralized national government. The Western provinces have strong separatist leanings, and leading provincial and national politicians openly espouse breaking off from the rest of Canada or joining the U.S. The Maritime Provinces also talk of joining the U.S.; Newfoundland, with its new-found oil potential, is talking of leaving the Canadian Confederation it joined in 1949. The Inuit of the Northwest Territories want autonomy for themselves, the way their neighbor, Greenland, has won autonomy from Denmark. Northern Ontario talks of separating from Southern Ontario, and the northern territories want provincial status in order to be freed from federal rule.

Rigidity in the face of regional aspirations increases the chances of a crack in Confederation; recognition of the relationship between democracy and decentralization, understanding that the stronger the parts the stronger the whole, adds to the reasons for the country staying together. The kind of sovereignty-association first proposed for Quebec has already gone through several transformations and could well evolve into the model for a revitalized country. The process under which this will happen can only be valid if the public is intimately involved at each step of the way. The decision is too important to be left to the whims of politicians. Except in case of war or national emergency, where a strong executive power is needed to respond quickly, no major decision need be made in a democracy without the benefit of direct public participation.

For referendum campaigns to be valid and informed expressions of the public will, a broad-based educational capacity is needed to disseminate the information and deliberate over its implications. It is no coincidence that René Lévesque's separatist side was strengthened by the grassroots network of non-political OUI and NON committees that spread across Quebec. Operating much like any other public interest group, 10,000 of these decentralized units within a matter of weeks had involved whole communities, convinced citizens to take sides, to debate, to lobby, to educate, to modify positions, to generate research—to breathe a little democracy into the society.

With both sides having access to information, propaganda was discarded as it soon discredited anyone who used it.

It's also no coincidence that the inept political leadership of the winning NON side was rescued by a grassroots organization of another kind—the "Yvette" movement of anti-separatist women that sprang up spontaneously to counter the separatist claim that federalism and feminism were incompatible.

The final verdict was one everybody could live with. Eighty-four percent of the public cast a ballot on voting day, leading Premier

Lévesque to admit that the Quebec people had "clearly given federalism another chance."

Then, as quickly as the referendum network had been set up before the referendum, it was disbanded after it. At low cost and great speed the most pressing social issue in Quebec took a giant stride towards a resolution.

While exclusively in the hands of government, the issue of French-English relations was used primarily to score political points. After 100 years of conflict the matter is being put into the hands of the people and it is finally being handled with dignity and courage.

We are finding that we humans are less fearsome than we feared, when governments aren't standing in the wings saying "let's you and him fight." Though man may never be a model of moderation, there was far less violence at the height of the unregulated Wild West than is present in the highly regulated, highly policed metropolises of today, and if ever we hope to live in any kind of harmony with each other the barriers between us must be made penetrable.

No society that is faithless to its basic premises can survive; not a socialist society, not a capitalistic society, not a democratic society. Viewed favorably, our governments today are proxy democracies, inherited from centuries ago when poor communication and great transportation distances foreclosed greater say by citizens in their government affairs. Viewed cynically, our governments are pseudo democracies, cloaks for centralized rule offering only slightly more pretence of choice than is presented to the Russian electorate on voting day. If we are to accept democracy—rule by the people rather than rule by an elite—we need less of a proxy and more of a direct democracy, less lip service to democratic principles and more exercising of those principles, less tampering with our instinctive organizational methods and fewer of the encumbrances that dominate our social and economic lives.

# 22

## *The Freed Enterprise System*

ONE OF THE last bastions of free enterprise left in the world is found in the USSR. In the markets of Samarkand, the mountains of Georgia, the fields of Siberia and the towns of Azerbaijan, the free market successfully survives next to the Soviet state monopolies, despite decades of purges and vilifications. There are 8,000 farmers' markets alone, thirty of them in Moscow, selling tens of billions of rubles worth of privately produced food—the produce grown in the peasants' modest private gardens; the slaughtered calves, chickens, and rabbits that have been privately raised. It is now legally a free market, quietly accorded its place alongside its powerful socialist partner.

In agriculture, the Soviets concentrate most of their efforts in the huge state farms, where the farmers are considered employees of the state, paid on a wage basis, and in the huge collective farms, where the farmers are considered members of the collective, receiving a share of profits in return for their labors. In addition, farmers on the collective are entitled to a tiny plot for their own use—these plots in total amount to 20 million acres or less than 1 percent of all the agricultural lands in the USSR. Yet from these plots come 62 percent of all Russian potatoes and 32 percent of all other fruits and vegetables; 47 percent of all Russian eggs; 34 percent of all Russian meat and milk; in all 27 percent of the entire agricultural output of the country.

The USSR has traditionally been a great agricultural power, once called "the bread-basket of Europe." Before the Russian Revolution in 1917, the port of Odessa was built to handle huge wheat exports. Today it is used to import grain to feed the Soviet people. All the farm machinery, fertilizer, and human fodder the state feeds to its large-scale state and collective farms has so far failed to raise production to the levels the state planners had forecast.

It isn't as if the Soviets are technological slouches. Their Sputniks were first in space; and in nuclear weaponry, that most developed of arts, they take a back seat to no one. These successes are due to their superb, centralized economy, the most centralized economy in the world. A decentralized economy would not have put a man into space for them or been able to concentrate enough men and materials to create a nuclear strike force. For these highly centralized purposes, decentralized means would have been ineffective, a poor match for the limited objectives of the work to be accomplished.

Americans, too, have put men into orbit and developed a powerful nuclear arsenal using centralized systems. Centralized systems are ideally suited to get one thing done in one place for one purpose using unlimited resources.

But to accomplish many things (such as growing many different crops and raising many different kinds of livestock) in many places (say across a huge country, where growing seasons, geography, and climate differ radically) for many purposes (from foods to feed people and cattle, to cotton for spinning looms, to pelts for tanneries) and using very limited resources that have to be carefully allocated (whether by price or by planners), only decentralized systems can be efficient. Whenever there are many different markets that have to be met, many different approaches must be used. Unable to get around that conclusion, centralized systems have changed the premise: there was no reason humans had to demand greater diversity than centralized systems could supply, so the centralized system systematically saw to the elimination of substantive differences between markets.

In North America, this was accomplished mainly through monopolization aided by the use of advertising, which taught us to dress, house, and feed ourselves just like everyone else. Mass markets were created in direct relation to the speed with which we abandoned our own personal creativity. In Russia, where crude devices such as advertising are abhorred, the markets were changed by fiat: the state decided the people should eat cucumbers when the cucumber crop comes in, tomatoes when the tomato crop is harvested. Although both can be grown on the same farms at the same time, the machines are most efficient when they are grown in rotation. The Russians have adapted effortlessly to their streamlined society: with the exception of a handful of the large Soviet cities, (where the residents are not as adaptable), throughout the USSR vegetables are ordinarily available only in season, and one at a time.

Free markets can exist only in decentralized systems but they are not easily wiped out. When they are outlawed, they re-emerge as black markets. Free markets are organic, common to every country in the world because they represent the way human beings naturally tend to organize themselves. Free markets are still in evidence in every village of Africa, Asia, and South America, and are making a comeback even in the highly centralized Western economies.

The damage caused by centralization is becoming increasingly difficult to sustain in all democracies with the possible exception of Japan, whose people have been culturally conditioned to accept the severest demands made upon them by the state.

Since the Second World War, Britain has consistently sought scale and integration in her businesses. Today she has a greater ratio of large, integrated and monopolistic organizations to small, independently-organized operations than any of her European trading partners and competitors.

The British disease is centralization, not the labor unions, which are just one of its many symptoms. Centralization has as a feature the need for specialization—maintaining large systems requires large support networks to avoid supply disruptions or distribution failures. To cut off a significant amount of the world's oil, as the PLO has made clear, would require sinking one oil freighter in the narrow strait leading out of the Persian Gulf, blocking the passage to all vessels. To disrupt an electrical grid only one link in the system needs to fail.

When the electrical grid is a nuclear one, the disruption can come not only from a downed transmission line but from a valve that sticks or, as almost happened in Ontario in 1979, by a strike of the utility workers: unlike the less-centralized hydro-electric systems or coal-fired systems, nuclear systems are too demanding to be kept operational by management personnel during a labor dispute.

In Montreal, a strike of elevator repairmen shut down office towers because of the possible consequences of an elevator malfunction. In France, the French government admitted "a fundamental problem" with its energy system after a nationwide power failure was caused by a sudden jump in consumption due to unusually cold weather. Thousands were stranded in subways and elevators, surgery took place by flashlight, traffic jammed and chaos ensued as the total blackout lasted two hours and fifteen minutes. Ten percent of the country remained in the dark after eight hours and two western areas (served by nuclear reactors) were affected for thirty-six hours.

When Britain's coal miners went on strike in 1972, the country's so highly interdependent economy was forced into a three-day work week. Had Britain already fully converted to nuclear power, a strike of the electrical workers would have given them the zero-day work week.

In any field—energy or education or elevator repair—centralization gives any specialist the power to hold the organization (or even society) to ransom by withdrawing his services. Everyone holds a trump in the house of cards we've constructed.

Unions, especially, have benefitted from this power, learning well each lesson taught them by the corporations they've organized against. Multinational unions were an inevitable consequence of multinational corporations; the secondary boycott, such as the one organized against Ford and CBC in support of the United Auto Workers strike against Ontario Blue Cross, was a tactic used a century ago by Rockefeller in securing his own monopoly. As industry became thoroughly interlocked, unions became thoroughly interlocked. The protection against exploitation that members of unions gained, like the protection against competition the corporations gained, came at the cost of security for all of society, which more and more is caught as an innocent third party in labor disputes between big companies and big unions. Our interdependent world economy, so praised by our politicians, is a euphemism for an insecure economy—a rationale for being dependent on events in Iran and Afghanistan or a fragile home front for our own welfare.

Seen in this light, Japan becomes the most vulnerable industrialized society of all, winning its present economic improvements not only at the sacrifice of individual diversity but potentially at the sacrifice of the Japanese homeland itself. From overwhelming dependence on foreign oil Japan is moving toward overwhelming dependence on nuclear power, putting all its eggs in a basket that will have to be held in a tighter and tighter grip. The long lead-times needed for Japan's centrally planned economy are forcing it to always think far in advance, even to the point of now considering the phase-out of its small-car automobile giants for future information-based technologies. While such far-sightedness is normally laudable, when an entire economy can collapse because of one wrong or premature decision it is not flexibility that is demonstrated but fanaticism. The same kind of fanaticism—in reverse—is being shown in North America by governments' decision to bail out the management of Chrysler, with those of Ford and General Motors to follow.

Japan or North America, or both, will have guessed wrong, with serious consequences for society. Wrong guesses can never be avoided, but their large consequences can. We could move toward a smaller-scale, independent economy, one based on cultural and geographic differences whose natural markets can best be matched by decentralized systems. As a result of the OPEC oil crisis, these markets have become impossible to ignore. Their growth represents the growth of the non-monopolized economy. The innovations that are developed to meet special, localized needs will then find applications world wide, increasing the world's store of useful, market-determined technologies.

Nothern Telecom has succeeded in a communication field dominated by multinational giants by basing its expertise on Canada's need to communicate over great distances between sparsely populated areas.

Farm-implement manufacturers on the Canadian Prairies have built up a half-billion dollar business based on the specialized needs of local farmers. When foreign manufacturers and large Eastern Canadian concerns proved ill-equipped to make production adjustments to fit their equipment to different terrain, smaller companies took over and profitably met the specific needs of specific Prairie farming districts. Similar needs in the United States and Europe were then discovered, and because most Prairie-made equipment is suited to dry-land farming, applications have been found for countries with similar growing conditions such as Algeria, Brazil, Mexico, Nigeria, and the Sudan.

The success of the Prairie manufacturers would not have been possible without a new information-sharing capacity developed to give manufacturers and parts suppliers access to knowledge of what local markets needed to be met. Information is often all that's needed for a manufacturer to risk new ventures, or a local parts supplier to displace a distant one.

With the advent of new information technologies, countless untapped small and specialized markets will become known to us—many through the TV screen. Torontonians can already press a button and discover the temperature, the specials at the supermarket, the arrival and departure times of planes, stock market quotations, hockey scores, lottery results, and community events.

Families in parts of Coral Gables, Florida can scan a list of movies, order catered meals, and do their banking via a telephone video screen. In Albany, N.Y., the contents of forty telephone directories are available to some families and businesses at the touch of a button. In Alberta and Texas computerized home systems are providing communications with police, fire and ambulance, including automatic fire alarms, remote utility reading devices, energy-consumption monitors and cold detectors (to avoid freezing pipes). Far more significant, however, is Canada's new Telidon system—a two-way TV device able to provide users with 10,000 "pages" of information on almost any conceivable subject, including information provided by the users themselves.

Invariably, access to information will lead to the elimination of the large centralized institutions that have effectively exercised information monopolies. This in turn will lead to the rejection of those outdated and expensive middlemen; we will make our own bargains when we choose to.

In every block, in every apartment building, people are taking out loans from the neighborhood bank while their neighbors are putting money into the same bank for the interest they earn. Often they shop around before making their banking decisions; often they go to considerable trouble arranging their transactions around the bank's hours and the loan officer's convenience. The bank's profits from their activities will help build the next branch for the bank while paying for all the paper and personnel it keeps employed.

When all the obfuscations surrounding these events are discarded, what can be seen is one neighbor lending money to another, and both paying a bank for the privilege. Information technologies are now available to put these people together through the two-way TV screen or any other information medium individuals might choose. The loans can be insured directly with insurance companies and the interest the banks would have claimed can be kept by the individuals. The money that goes into bank buildings can be spent more productively elsewhere.

There need be no fear of insurance companies unduly attaining some of the power lost by the banks. In a free market, with insurance company dealings open to everyone's scrutiny, competition would regulate the size of insurers.

We are learning that there are no large natural monopolies, only monopolies that are self proclaimed for self-evident reasons. In practice, the monopoly function becomes that of a conduit—whether as a streetcar track or electricity transmission line or telephone cable—tolerated to avoid duplication but a monopoly for no other purpose. This conduit function has been parlayed by public and private organizations into monstrous monopolies which are only now being questioned. The telephone company's control over everything touching its telephone wires continues to be challenged, and it is continuing to lose its monopoly over information being transmitted through its cables. Privately owned computer data, electronic mail, telex, cable TV, and other innovations can be transmitted through wires connected to privately owned Mickey Mouse phones and computer terminals. The phone company need charge only on the basis of the information being carried along its system without the right to impose monopoly censorship over data or monopoly foreclosure of competing terminals. Electrical monopolies can be similarly cut down to size if they become only a conduit of electricity—carrier for any public or private electricity generator and electricity purchaser on its system. The buyers and sellers of electricity (whether publicly or privately owned) will be able to conclude their negotiations in a free market with far less waste than a giant utility generates. They will be able to make consumer purchases from this coal station or that hydro plant based on their own needs rather than the utility's wants.

This straightforward business practice was envisaged and adopted by the proponents of Ontario Hydro at the turn of the century—the utility was created to be an intercity distributor of electricity at the service of the cities. Through Hydro's transmission lines, municipal (and industry-owned) generating stations sold their surpluses to each other. The municipalities had clear control of their power production and the public—through plebiscites— had clear control over the building of new electricity plants. As an early corporate history of Ontario Hydro spelled out, Hydro's function was that of a conduit: "If the people of a municipality desire to obtain a supply of power from [Hydro], a vote is taken at the polls and if the result is favorable . . . the municipality is empowered to make a contract with [Hydro] for the amount of power required." Although Ontario Hydro would build and finance the municipal system, "the municipality repays the money out of earnings over a period of thirty years . . . at the end of which time the entire plant and equipment will belong to the municipality; thus the people will eventually be the owners of the whole undertaking. The basic principle of the whole scheme is a partnership of municipalities formed to obtain power at cost, each municipality paying for the cost of service received."

Even Niagara Falls power was privately generated—the province of Ontario earning royalties from the water rights leased to industry. Hydro did not see itself as a major producer of power until 1913, when a scheme was first laid before the provincial government requesting Hydro have "its own development works [to] generate power on a large scale." Not until 1917 was Hydro able to deny municipal voters a direct say in its expansion plans.

Had the original type of electricity system not been subverted by the power Ontario Hydro was later given, and had the municipalities determined for themselves how much electricity to produce and sell, a highly efficient system would have resulted. Had the municipalities decided to sell the electricity at its market value to allow the private electricity producers to compete in a free market, the system would have had an unparalleled efficiency.

Beginning in 1980, studies in Canada and the U.S. started uncovering the enormity of the public sacrifice made to sustain electrical monopolies. According to the Mellon Institute, American utilities produce almost four times too much electricity because they operate outside the rules of the free market. According to Energy Probe, as much as two-thirds of Ontario's electricity "needs" would vanish if electricity were priced at its true cost. The implications are staggering, and go far beyond financial considerations. In Ontario's case, all nuclear and coal plants become anachronisms; the province more than meets its real needs through hydro-electric power. For the United States, all nuclear, all oil plants, and most coal plants become obsolete as existing waterpower can meet more than half the country's consumption. The balance can come from coal or any number of cleaner options, including windpower, which NASA has concluded can approach the level of America's hydro-electric production.

But the financial considerations are also no trifling matter, not for electrical utilities or for transit utilities.

Assets of hundreds of billions of dollars could be returned to private ownership and private control where, because of the newfound efficiencies, they will be diverted to more vital uses than excess generation. The public-transit utilities which collect fares from their passengers can leave that function to whatever public and private vehicles want to compete to use their tracks or roadways, charging only by distance traveled and costs incurred in scheduling. This would lead to numerous innovations to meet off-peak and irregular needs better than the rigid and all-too-often poor service presently available.

Seen for what it is, the concept of the monopoly becomes in practice irrelevant, one of degree rather than fixed definition. A temporary monopoly is all mass transit outfits in Manila can claim, with routes annually auctioned off to private firms, providing the municipality with a profit and the passenger with a level of service (if not safety) unseen in North American cities. The monopoly becomes comparable to any other contractual arrangement, with new bidders allowed to compete on price and service.

The equity and efficiency of the free market envisaged by Adam Smith comes from the availability to the consumer of all the information needed for his decision: right there, in stall beside stall, the purchaser could evaluate the tomatoes he wanted, note texture, size, color, ripeness, and shape literally alongside all other tomatoes on the market. He could not be easily fooled into overpaying or into buying something he didn't want. Then came industrialization, and life became complicated. Cereal boxes became larger than their contents, cars could rust after a single salt season, products broke down with planned abandon. All became guesswork with the worst-informed consumers being those most highly prized. Information was inaccessible; a free market was inconceivable.

New communications technology has now made it possible for information of any kind once again to be transmitted at low cost to any place. It is now possible to have an instant electronic catalog of all goods, to know what and who made each and how long they'll last and how much they'll cost to service and how much to fuel. It is possible to once more have a functioning free market, with access to all goods and services in the world through the movement of information—the greatest energy conserver and decentralizer of all.

The goal is a desirable one.

The market, after all, is the original, organic form of economic organization still practiced in the villages of the Third World with flawless equity. That it is practiced less equitably in the industrialized nations is not an indication that the free market should be abolished but proof that we must return to its basic principles if we are to restore it. In the local market of any Third World town or village, the consumer has all the information about his purchases literally at his fingertips, and is capable of making the best possible decision for himself. It is the lack of this information, above all else, that has undermined the competitive marketplace for the West. And it is the restoration of information, above all else, that is a necessary precondition to a revival of free-market mechanisms.

# 23

## *The Unwritten Agenda*

POOR ONTARIO HYDRO. It tries so hard to belong. To avoid criticism for the inefficiency of its nuclear reactors and to produce something wholesome for the people of Ontario, the giant utility put its reactors' warm waste water to work growing tomatoes in nearby experimental greenhouses. The greenhouses were to blossom into 150-acre "agriparks" growing local produce to displace imports from California and Mexico.

With the discovery that the tomatoes contained unusual amounts of radioactivity, including ten times normal levels of tritium—a carcinogenic element also feared capable of causing genetic defects—Hydro's do-gooding was at an end. Now it would proceed only over the public's dead bodies.

Those poor multinational electricity corporations. To solve the problem of America's dependence on fuels shipped halfway around the world, General Electric, Westinghouse, and others decided to meet U.S. energy needs by building solar satellites in space. Benefitting from twenty-four-hour unobstructed sunlight, each satellite would beam to earth stations (by microwave) enough electricity for 5 million people.

The zeal with which this $88 billion program was clearing legislative hurdles in an American congress eager to avoid being held hostage over energy was dampened in 1980 by an intelligence report from the CIA indicating the Soviet Union has developed a ground-based laser weapon capable of destroying space satellites.

Poor Egypt. To increase the productivity of the Nile Valley, control flooding, preserve water for use in dry years, reclaim 800,000 acres of land for farming, and generate 6 billion kilowatt hours of electricity per year, the Egyptians called on the Russians to build them the Aswan Dam—seventeen times bigger than the Great Pyramid. Aswan did these things . . . and more.

Because the dam cut off the annual flood of 30 billion tons of silt (which used to make the Nile Valley the world's richest), the fertility of Egypt's soil has declined, farmers now buy fertilizer to replace the fertile mud, the banks of the Nile are eroding, and parts of the Mediterranean coast are eaten away. With their food chain broken, fish fled from the river's mouth and the sardine industry—18,000 tons per year—has been wiped out. Disease is spreading as the waterborne bilharzia parasite now affects half the population with intestinal and urinary ailments. Satellite photos show that—without the annual river flood coating the land with mud—massive dunes from the desert's Sea of Sands are drifting over the fringes of the farmland. Perhaps most crucial, until drainage is installed in the irrigated lands, salt rising to the surface in white crusts threatens to turn farms to wasteland. Serious thought is now being given to boring holes through the $1 billion disaster and restoring the old ecological (and economic) system.

The results were predictable and, in fact, predicted. But faith in technology has often been the refuge of those practicing bad science or overriding common sense. The multinationals' space project does not need an enemy laser to appear preposterous. Each receiving antenna on Earth would need over thirty-six square miles of land. Should the earth-bound microwaves be transmitted off-target by only a fraction of a degree, the potential of the high-powered beams could destroy a city. Should a malfunction in space require a service call from Earth it may take time delivering the proper parts via Cape Canaveral.

Yet even with the existence of Soviet lasers, it is not difficult to imagine the corporate lobby convincing congress that the satellite's defence can be secured with new weapons technologies—we would then be arming space as well as Earth, escalating the arms race along extra-terrestrial supply lines, and all to become more vulnerable than before.

It would also not be difficult to imagine Ontario Hydro proceeding with greenhouse vegetable production after establishing "safe" levels of radioactivity for public consumption. Should the new industry survive public opposition for its first few years and prosper, the public would have mixed feelings about closing it down even should cancer rates and birth abnormalities start to rise. Compared to the annual deaths and disabilities from the automobile or tobacco, the additional toll from radioactive tomatoes would indeed be "a miniscule amount." Rather than upset a local industry, throw people out of work, and hurt the region's balance-of-payments situation, the public might well allow continued greenhouse production in the hopes this hazard could be minimized, the same way we are presently allowing our air and drinking water to bear an ever larger radioactive burden.

The acceptance of radioactive tomatoes in Ontario would be paralleled by the use of nuclear waste heat in other industries, enlarging the numbers of those with a vested interest in nuclear power. Imperceptibly, radiation-related deaths would climb. Because we are reasonable beings, we might even decide that the same standards found acceptable for the automobile—50,000 deaths per year—should also be applied to the nuclear industry. This logic, propounded by the nuclear industry, has also attracted others with a grim sense of fair play. We would then have developed a limit to growth different from those previously anticipated. At the rate of 50,000 unnatural deaths per great technological benefit, there would be only so many great benefits we could afford on a continent whose population growth has all but stopped. New technologies could now be sanctioned only as long as the presence of human fodder were maintained. Human masses would explicitly exist (or perhaps be produced) to satisfy the needs of machines; brute force would have replaced human ingenuity as mankind's basic means of survival.

Solar satellites, mammoth dams, and nuclear power plants, all highly centralized methods of providing energy, could only be seriously contemplated by highly centralized and non-democratic institutions like the private electrical monopolies (GE and others in the case of solar satellites), public monopolies (Ontario Hydro and others in the case of nuclear power), and state monopolies (the USSR in the case of the Aswan Dam). They are beyond the scope of individuals and small-scale institutions, which invariably find more appropriately scaled solutions to problems, even to large-scale energy crises.

Due to a shortage of wood 2,500 years ago, the ancients designed and sited houses to obtain the most warmth from the sun. The Greek playwright Aeschylus wrote that civilized people, unlike barbarians, had houses that "turned toward the sun." Pliny the Younger boasted that his villa north of Rome was warm because it "collects and increases the heat of the sun" since its windows were placed "to catch the low winter sun." Vitruvius, a Roman architect, said, "we must design houses according to climates" and urged buildings "be closed on the north side with the main living sections facing the warmer south side." The city of Olynthius in northern Greece, at the same latitude as New York or Chicago and with a population of 30,000, was the first passive solar city. Main rooms were on the north sides of the courtyards, facing and opening to the south. Houses had two storeys, but the south wings were kept to one storey so as not to block the sun's rays from entering the main living rooms. In ancient Rome, erection of a structure that blocked a neighbor's sunlight was illegal, a prohibition maintained in the Western World until the turn of the century, when progress took a new turn.

Energy technology—instead of being refined to augment the arts of the ancients—was used by the new energy monopolies to replace them. Further refinements would not take place until the OPEC oil crisis signalled the end of the petroleum interlude. The refinements can now be seen across the continent at the individual, community, and municipal levels; they are vying against the senior levels of government and large corporate interests promoting centralized systems for control of the new energy era before us.

The cradle of democracy did not have to be in Greece, but the energy independence the Greek citizen achieved was a necessary precondition for Greek democracy and a contributing factor to Greece's Golden Age. To other societies, which communally gathered food and fuel for their survival, individual freedoms could have no meaning.

In twentieth century form, a choice between comparable societies awaits us. Either communal, centralized systems like nuclear power, or individualist, decentralized systems based on solar energy. Human living patterns for the next millenium will have been firmly established by the turn of the century, and they will be established based on the energy decisions we make today.

# ACKNOWLEDGMENTS

As disconcerting as it is for an author, to turn chapters over to others and have them dissect, rearrange, rewrite and reject whole sections at a time—all the while demonstrating a far greater understanding of the manuscript's goals than its author—tradition requires their contribution be credited here.

Alex Kisin several times changed the direction of this book, convincing me to write additional chapters and discard redundant ones. Herb Stovel caught numerous inconsistencies in my arguments while doing his best to restrain my penchant for rhetoric and verbosity. Peter Marmorek—whose usefulness as a critic was limited by his intuitive understanding of decentralization—merely rewrote awkward passages, and straightened me out when I got my facts wrong, and spared me a good deal of embarrassment.

Evidence of their inability to fully cope with their task can be seen from the many improvements later made to the text by Darla Rhyne and Norman Rubin. Both, however, doubt they have done enough to salvage it.

For my part, I welcomed all changes with ambivalence.

Many people aided me in research. Marilyn Aarons was especially determined in digging out facts I'd long despaired of finding. My appreciation as well goes to Moni Campbell, Gordon Edwards, Bill Glenn, Willie Kisin, Jack McGinnis, JoAnn Opperman, and Ralph Torrie, and to Jack Gibbons for providing an economist's perspective. About a dozen lawyers wrestled with my chapter on monopolies, which I believe will prove the most controversial in this book. The legal views were on the whole refreshingly diverse and surprisingly emotional. All contributed to a refinement of my arguments (not to mention some eradication of my errors). Among those whose comments changed the chapter (never, I should say, to the point any fully endorsed all its premises or its implications) were Don Jamieson, Andy Roman, René Sorell, John Swaigan, Robert Timberg, and Ken Wolfson.

I must also acknowledge the dedication of Dawn Aarons, Joan Burke, Bill Grove, John Ridelle, and Grace West in volunteering their time in typing and proofreading the several drafts.

Information provided by public interest groups was indispensable. I am deeply appreciative of the co-operation of organizations like the Canadian Coalition for Nuclear Responsibility (CCNR) and the Ontario Public Interest Research Group (OPIRG) for having generated a wealth of otherwise unavailable information.

Special thanks to Stephen Aarons for so graciously turning my mountains into molehills, to Mr. Robert K. Arnold for teaching me to look at world events from a longer and broader perspective, and to David Brooks, whose casual observation that "because energy flows through all systems it necessarily influences all systems" led to the train of thought that resulted in this book.

*Energy Shock* contains observations, opinions, and expressions of political philosophy that are personal and do not necessarily reflect the views of Energy Probe or any of those kind enough to have helped me in my work.

# SOURCES, NOTES AND SELECTED READINGS

## 1 *World in Transition*

Nuclear power's 1.3 percent is measured in terms of secondary consumption—not the gross amount of energy produced in the nuclear reactors (most of which is unusable) but the net amount which actually reaches the consumer. A full breakdown of secondary energy use in Canada is available in Norman Rubin's *What Keeps Us from Freezing in the Dark* (Energy Probe, 1980).

Those looking for other examples over-ebullient nuclear proponents should see *Accidents Will Happen: The Case Against Nuclear Power,* published by the Environmental Action Foundation, or any early work written by Edward Teller.

## 2 *How Volatile, Energy*

Of the $300 billion that the Royal Bank's chief economist, Ralph Sultan, sees Canada investing, almost $200 billion is for electricity expansion—an eventuality that is difficult to imagine with the tremendous oversupply of generating capacity across Canada. My guess is that even the $100 billion for non-electrical expansion will not be spent as demand for all fuels tapers off due to the economics of conservation. In the United States, a study by Gyftopolous and Widmer (reported in *Techonology Review,* June 1977) estimates that the capital costs of a conservation program that would maintain an average growth in GNP of three percent per year would be about one-quarter the cost of financing an equal growth without conservation.

### 3 *Energy Ideologies*

Center for the industry's impressive public relations campaign was the "University of Light," G.E.'s suburban headquarters at NELA park, Ohio. Unlike the inefficient electric system in place in North America today, G.E.'s original installation there was a marvel of good design: a district heating system pumped the electric plant's waste steam to outlying buildings, the pipes having been thoughtfully laid under sidewalks to keep walks free from snow in winter.

More information on the controversy between Edison and Westinghouse over A.C. and D.C. can be found in the *IEEE Proceedings,* (Vol. 6, No. 9, Sept. 1976). The energy saving of having D.C. systems is difficult to predict, but they may be far more profound than anticipated. Joe Umanetz, a teacher of conservation technologies, manages quite comfortably in his country home near Durham, Ontario, on 1,000 windmill-generated watts. His D.C. television set requires only twenty-four watts, his D.C. stereo less than one watt. Without the wear of A.C.'s oscillations, his lightbulbs have yet to burn out. Because he generates his electricity locally, there are no significant line losses, and of course, D.C. systems do not need transformers.

G.E.'s skill at public relations can best be explained by itself: see *Lamps for a Brighter America* by Paul W. Keating for a fuller and more sympathetic view of the electrical industry's marketing activities.

### 4 *Power Brokers*

Before you think too ill of John B. Sheridan, whose diabolical public relations genius did so much to convert us to the electrical faith, be heartened to know that in later life his conscience got the better of him. His colleagues had no similar change of heart, however, and refused to let him undo the damage he had perpetrated. After much frustration, Sheridan took his own life.

Most of the information dealing with utility public relations can be found in *The Public Pays: A Study in Power Propaganda* by Ernest Gruening, who abstracted the testimony and findings of a three-year Federal Trade Commission study supplementing an investigation by a committee of the U.S. Senate. More on G.E., and other cartels, can be found in Wendell Berge's *Cartels—Challenge to a Free World* and in maverick G.E. vice president Theodore Quinn's own books, especially *Unconscious Public Enemies.*

*The Great Pierpont Morgan* by Allen Frederick provides details of Morgan's relationship with Edison and *The Age of Giant Corporations* by Robert Stobel provides details of Morgan's interconnected electrical empire.

**5 *The Socialist Menace to the North***

Ontario Hydro's ambitious plans for nuclear power were incorporated into the federal department of Energy, Mines and Resources' major policy statement, *An Energy Policy for Canada*—Phase I (1973). On page 101, Vol. 1, of this report, a graph charts the country's energy developments to the year 2050. Between 2040 and 2050 Canada was expected to have built the equivalent of a 500 megawatt reactor (the size of the Pickering reactors) every 1.4 days. Ontario Hydro would no doubt have been incensed in 1973 at being allocated only one-quarter of the country's nuclear reactors, but on this modest assumption of their immodest dreams, the effect would have been one reactor built in the province every five-and-a-half days. Of course, neither Ontario Hydro nor Energy, Mines and Resources ever thought through the full import of their projections. It took Ralph Torrie, the author of *Half-Life* and one of Canada's most devastating nuclear critics, to make the calculation that showed exactly what our authorities were countenancing.

Several histories of Ontario Hydro have been written, the best by far being Plewman's *Adam Beck and the Ontario Hydro.* Written by a municipal alderman, the subtleties of the nature of the Ontario Hydro struggle were well understood. *The People's Power* by Merrill Denison provides a more recent, if less critical work. To fully understand the public vs. private conflict, William E. Mosher's *Electrical Utilities: The Crisis in Public Control* is recommended. The Sutherland and Gregory royal commission reports make surprisingly pleasant reading, as does the uncritical harangue by Professor Mavor in his 1916 series of articles in the *Financial Post.* Another harangue by Reginald Pelham Bolton, *An Expensive Experiment, The Hydro Electric Power Commission of Ontario,* is less skillful, despite having greatly influenced New York State against public ownership.

For a fine analysis of what went wrong with co-ops in the United States, Alden Meyer's *Lines Across the Land* nicely dissects the rural electric co-op (and with it the co-op panaceas, so often espoused as economic remedies).

"Technology and Public Power: The Failure of Giant Power" by Thomas Parke Hughes, published in the *IEEE Proceedings* (Sept. 1976) gives more background on the role of Governor Pinchot and Giant Power.

Additional details of nuclear industry intimidation (and vivid elaborations of the examples given) can be found in *Cases of Misinformation and Attempted Suppression* by Dr. Gordon Edwards, an exhibit listed in the Select Committee on Ontario Hydro Affairs' *Interim Report on Nuclear Safety* (Dec. 1979). The graduate student accosted at the University of British Columbia, incidentally, was Dr. Edwards himself, now the chairman of the Canadian Coalition for Nuclear Responsibility and the inspirational leader of the anti-nuclear movement in Canada.

### 6 *Monopoly Money*

The biography by Collier and Horowitz, *The Rockefellers* and the devastating turn-of-the-century exposé by Ida Tarbell, particularly her *History of the Standard Oil Company,* provide the basis of the historical data found here. Henry Demarest Lloyd's vivid description of Standard's promiscuous political activities ("The Standard had done everything to the Pennsylvania Legislature except refine it") comes from his *Wealth Against Commonwealth,* while the interlude discussing Rockefeller relations with the Russian oil kings was reported by *Bradstreets* in 1893. The power of the Rockefeller and Morgan organizations was brilliantly illuminated by John Moody's 1904 classic, *The Truth About Trusts* ("These two mammoth groups jointly . . . constitute the heart of the business and commercial life of the nation, the others all being the arteries which permeate in a thousand ways our whole national life, making their influence felt in every home and hamlet, yet all connected with and dependent on this great central source, the influence and policy of which dominates them all").

### 7 *Fuel Wars*

The multinationals' role in World War II is documented in many places, a surprising amount of damning evidence having been available for decades, even in *New Horizons,* Standard of New Jersey's own corporate history (although a little reading between the lines here is required). For a straightforward presentation of the I.G. role, Borkin's *Crime and Punishment of I. G. Farben* is highly recommended. Details of Ford's and General Motors' wartime activities abroad are available in "Strategic Bombing II," a U.S. military document declassified in 1973.

The only popular description of the planned corporate coup of America that I've found lies in *DuPont—Behind the Nylon Curtain,* a history of the DuPonts that painstakingly describes one of the greatest American fortunes founded on fuels. Post World War II material can be found in *The Rockefellers* (for information on David Rockefeller and The Chase) *Multinational Corporations and U.S. Foreign Policy*—Part 4: 1974 (for a U.S. Congressional view of oil company influence in the Middle East) and *Continental Corporate Power* and *Modernization and the Canadian State* (for the Canadian perspective).

### 8 *Monopolies: Building Them Up, Breaking Them Down*

I cannot direct you to any source that still questions the concept of the corporation as person although a century ago, during the period of its rapid evolution, the issue was highly contentious. Chief Justice Marshall, who is said to have remade the American constitution by his judicial reading of it, had earlier laid the foundation of the body of private corporate law with his classic statement: "A corporation is an artificial being, invisible, intangible and existing only in the contemplation of the law. Being the mere creature of the law, it possesses only those properties which the charter of its creation confers on it . . . These are such as are supposed best calculated to effect the object for which it was created . . . Among the most important are immortality, and, if the expression be allowed, individuality: properties by which a perpetual succession of many persons are considered the same, and may act as a single individual. They enable a corporation to manage its own affairs and to manage its property without the perplexing intricacies, the hazardous and endless necessity of perpetual conveyances for the purpose of transmitting it from hand to hand. It is chiefly for the purpose of clothing bodies of men, in succession, with these qualities and these capacities that corporations were invented and are in use. By these means, a perpetual succession of individuals are capable of acting for the promotion of the particular object, like an immortal being. But this being does not share in the civil government of this country, unless that be the purpose for which it was created. Its immortality no more confers on it political power, or a political character, than immortality would confer such power or character on a natural person. It is no more a state instrument than a natural person exercising the same powers would be."

Historical accounts of the corporation's evolution can be found in *Pierre S. DuPont and The Making of the Modern Corporation*, *The Supreme Court and American Capitalism*, *Public Policy and the Corporation*, and *The Corporation in American Politics*. For analyses of monopoly influence, a wealth of distracting books has been published. Particularly refreshing are *An Anti-Trust Primer*, *Cartels—Challenge to a Free World*, *Monopolies and Patents*, and *America, Inc.* An interesting counterweight to the Canadian Royal Commission on Corporate Concentration's report (1978) can be found in an 1969 report of the Federal Trade Commission that warned that the trends toward industrial centralization "pose a serious threat to America's democratic and social institutions by creating a degree of centralized private decision making that is incompatible with a free enterprise system, a system relying upon market forces to discipline private economic power."

For a critical view of the streetcar, try *The Selling of Rail Rapid Transit*, which shows some seamy sides of the streetcar interests. I also found Cornehls and Taebel's *The Political Economy of Urban Transportation* extremely useful, particularly because of the historical perspective in which it placed the automobile. For the beginnings of the modern defence of the automobile, see the *IEEE Spectrum* (Nov. 1977).

## 9 *Trafficking*

The airplane is not quite as inefficient as the car—120 passenger-miles per million BTU compared to 110. Railways and buses have the highest energy productivities: 630 and 340 passenger-miles per million BTU respectively. Other breakdowns of transportation vehicle efficiency are available in Barry Commoner's *The Poverty of Power.* Here it is also explained that prewar automobile engines, operating at low compression ratios and therefore at relatively low temperatures, did not produce enough nitrogen oxides in their exhausts to trigger the photochemical process that is responsible for smog.

G.M.'s destruction of public transit is brilliantly documented in *American Ground Transport* by Bradford C. Snell (printed in the U.S. Congress, Senate, The Industrial Re-Organization Act, Hearings before the Subcommitte on Antitrust and Monopoly of the Committee on the Judiciary, United States Senate, S. 1116, 93rd Congress, 2nd session). This report is worth whatever efforts you may require to obtain it. Unfortunately, it is not available through the U.S. Senate—the government, at General Motors' request, agreed to the unusual measure of withdrawing this material from circulation.

## 10 *On Track*

The Bradford Snell report *American Ground Transport* (see chapter 9) was relied upon extensively in this chapter, supplemented by articles in the *IEEE Proceedings* (Sept. 1976, pp. 1350-60), *Traffic Quarterly* (April 1977), *Canadian Business Review* (Summer 1977), and the *Upper Canada Railroad Society Newsletter* (Feb. and March 1962, April 1963, among others). G.M. apparently proposed bus trains as early as 1967. The idea was presented later as well by the Automobile Manufacturers Association's *The Potential for Bus Rapid Transit* (1970). The concept, as described there, is for 1, 450-unit bus trains, which in cities would operate in tunnels. *Forbes* (Nov. 15, 1970) contains some of the business euphoria over the motor home (more euphoria in *Fleet Owner,* July 1972).

*The Imperial Oil Review* (June 1959) describes the commitment of the Canadian railroads to diesel locomotives, but less persuasively than *Canadian Transport* (Oct. 1959), which breaks down expenditures on transportation in a manner which highlights the disproportionate nature of the expenditures: in the ten-year period 1948-57, the cost of the roads alone for servicing motor vehicle traffic approximated the entire capitalization of the Canadian railroads. Another look at auto subsidies in Canada is available in *Urban Transportation Development in 11 Canadian Metropolitan Area*—(1966).

A positive view of rail's chances was published by the American Public Transit Association (APTA) in a report called *The Case for Rail Transit,* which enumerates many of rail's recent successes and provides encouraging statistics of rail vs. road, but fails to properly address the issue of automobile subsidies (although APTA does cite a University of Iowa study that shows the cost of constructing a six-lane suburban highway with a capacity of 1800 vehicles per lane per hour to be 88 cents for each person per mile who uses it. In contrast, the Iowa report found the per-person cost of building a ten-mile rail segment with six stations and having a capacity of 18,000 persons per hour to be 22 cents).

The railroads had an unexpected early champion—as described in the *IEEE Proceedings* (Sept. 1976). Henry Ford acquired control of the Detroit, Toledo and Ironton Railroad and tried to revolutionize rail as he had the automotive industry. In 1929 he admitted defeat, Ford's venture is significant, however in discounting the existence of a nefarious plot, against rail by the auto giants. The same issue of the *IEEE Proceedings* is also useful in its disparagement of the diesel.

Canada's railway industry fared better than America's due to a policy of decreasing regulation in Canada, according to R. R. Latimer, vice president and senior executive officer of CNR, in an article for *Railway Age,* (May 28, 1979). Interestingly, it took the highly bureaucratized CNR and CPR seventeen years to take full advantage of their new ability to negotiate rates directly with customers—a process that occurs spontaneously in a free market. "Looking back, I would say that the process could have been speeded up," Latimer concluded. "Perhaps we could have done it in 8 to 10 years."

### 11 *What the Traffic Will Bear*

Streetcar buffs will enjoy *Remember Montreal's Streetcars, Montreal's Electric Streetcars,* and *Stampede City Streetcars.* For the minutest of details about streetcars (especially the TTC) I recommend just about every volume of the Upper Canada Railroad Newsletter which—though tedious to read—yields so much useful trivia that the nature and scope of the streetcar industry's battle becomes well defined. More on the TTC and transit in Canada can be found in *City Magazine* (Vol. 3, No. 4 and 5), *Transit Canada* (Vol. 14, No. 1 and 2) and *The Financial Post,* (March 4, 1978). Sources used regarding the subsidies offered to the car come largely from periodicals: *Financial Times* (Jan. 21, 1976); *The Economist* (Aug. 21, 1978); *Farm and Country,* (March 13, 1979); *The Financial Post* (March 4, 1978; Nov. 18, 1978); *Canadian Business* (Feb. 1979); *City Magazine* (Jan. 1977); *Globe and Mail* (Nov. 30 and Dec. 1. 1976, Jan. 25, 1977; Feb. 7,

Dec. 9 and Dec. 11, 1978; April 12, June 4 and June 5, 1979; among others). Re: Amtrak, see *Railway Age*, (Aug. 13, 1979).

*American Ground Transport* (see chapter 9), as well, provides much of the information in this chapter and the *Journal of Urban History* (May 1979) is highly recommended.

The $110 subsidy to suburban transit riders can be found in the 1975 *Metropolitan Toronto Transportation Planning Review* (Vol. 27, p. 79), quoting Research Report 7311 ("Income Distribution of Effects of Urban Transit Subsidies") published by the Department of Economics, University of Western Ontario. The subsidy has been stated by me in 1980 dollars.

## 12 *The End of Suburbia*

Many of my views of suburbia stem from two years of exile there. However, I found much support in the works of Lewis Mumford, particularly *The City in History* and *The Highway and the City*.

The works of Humphrey Carver (called by some the father of the Canadian suburb) also proved useful, including his article in the September 1978 issue of *City Magazine*.

John Sewell's article in the January 1977 *City Magazine* exhibits exceptionally clear thinking and his description of the rise of the Canadian suburb is the best I've seen. For less texture in your prose, try *Suburbia: Costs, Consequences and Alternatives* (Urban Studies Program, York University).

Levittown and the American suburb is exhaustively covered in *The Levittowners* and well summed up by Gertrude Stein's oft quoted "There's no there, there."

## 13 *Necropolis*

Mumford's *City in History*, provides unsettling parallels between our use of energy today and in the Industrial Revolution, and his descriptions of eighteenth century industrialization are paraphrased extensively in this chapter. To get the full flavor of the era, his Coketown section should be read ("humans crippled and killed almost as fast as if they'd been in the battlefield"). The nuclear study referred to is the Rasmussen study, commissioned by the Atomic Energy Commission and considered to have the best data available. My statement that the chance of a meltdown in the Toronto area is one in fifty (or 2 percent) is an understatement: according to the Royal Commission on Electric Power Planning, the best estimate is 3.6 percent and the estimate may be off by a factor of five, i.e. the chance of a meltdown could be as high as 18 percent.

### 14 *Journey Into the Center of the City*

The division of labor concept is also claimed by the USSR, and aptly so, as it well represents the Russian approach to labor. Discovered by a simple laborer named Stakhanov (according to Stalinist history), Stakhanovism became a symbol of socialistic progress. Those who treat this concept with reverence in the West might enjoy the reverence of their Soviet counterparts in Molotov's *What Is Stakhanovism?* (1935). For more planning visions of how well the auto would be integrated with the city see the May 1969 issue of the *Journal of Urban History.* Those who still advocate street widening as a solution to urban congestion and urban decay will take heart knowing their solutions have been employed since at least 1915, when Nelson Lewis, Chief Engineer for New York's Board of Estimate argued that "With better paving the traffic is becoming more diffused. The driver of the motor car naturally selects the more attractive streets. The more attention that is given to street details the better the character of the buildings fronting upon the streets, the more general the introduction of trees and shrubbing—all of which are receiving the attention of progressive real estate developers and those engaged in city planning work—the less will be the concentration of motor car pleasure traffic on certain streets."

### 15 *The New City State*

Detroit's rebirth will take a good deal more than its Renaissance Center, as is explained in *Urbanism—Past and Present* (Summer 1978). Using 1970 as a baseline, between 1970 and 1975 Detroit lost fully half of its manufacturing industry, with more to come. Next to Pittsburgh's the most specialized manufacturing economy among major U.S. metropolitan areas, Detroit lies at the center of the Oshawa-Cincinatti-Milwaukee automotive triangle which contains almost 90 percent of North American automobile employment. The decisions of both the Canadian and U.S. governments to prop up the auto industry (making the area ever more vulnerable) puts Detroit's rebirth even farther into the future. Descriptions of how cities do experience renaissance can be seen in Jane Jacobs' *The Economy of Cities* and *The Death and Life of Great American Cities;* and in Lewis Mumford's *The City in History.* A truly terrific little newspaper about urban agriculture is *City Farmer,* published in Vancouver. Many of my examples are derived from it (I only regret *City Farmer* does not have a longer history, that it might have provided me with more material).

The information on methane gas cogeneration systems comes from a variety of sources, but most usefully from Barry Commoner at Washington University in Missouri, whose concepts are pioneering the field of organic fuel technologies. Technical information on the FIAT cogenerator is available from FIAT itself. World Watch Paper no. 17, *Local Responses, to Global Problems* by Bruce Stokes, enlarges the self-sufficiency theme.

**16 *Beyond the Barricades***

Several chapters in my last book, *The Conserver Solution*, deal with the prospects for recycling (see especially "Enter The Recycler"). The bicycle data used was primarily obtained from *Traffic Quarterly* (Jan. 1977) and *Traffic Engineering* (March 1977) regarding the U.S.; from Barton-Aschman's City of Toronto Bikeway Study regarding Toronto. *Harrowsmith*'s article on the Chinese-run farms in British Columbia (No. 28, 1980) was brought to my attention in time to write it into this chapter as a North American example of a technique long known in other parts of the world.

The breakdown of regional government is only a matter of time, and I regret the limited space I dedicated to this theme. Examples of regional governments' failure are abundant and make for amusing reading. The survey cited was reported in *Urban Future Idea Exchange* (June 15, 1979).

**17 *Shoot the Planners***

The Toronto alderman who held firm against Hydro's intimidation started by asking questions in a field he knew little about. Now knowledgeable and one of the most daring thinkers in the energy field, Richard Gilbert spearheads the city's efforts to wrest control of energy from senior levels of government.

End-use planning techniques were popularized by Amory Lovins in his modern classic, *Soft Energy Paths*, and are perhaps the single most effective tool in countering conventional energy planners by showing how ludicrous their forecasts are. But when the critics begin to treat their critiques as optimum future blueprints, and produce rigid documents to aid in their implementation (as sometimes happens), I believe their usefulness is lost.

**18 *The Energy Toll***

Samuel Epstein's *The Politics of Cancer* and the Environmental Defense Fund's *Malignant Neglect* were my sources for most cancer-related material, and both books are highly recommended for their fresh perspectives on environmental cancers.

The notion that no synthetic chemical can be indefinitely isolated from the environment is strongly stated in the Don Chant/Ross Hall article in the *Probe Post* (Jan.-Feb. 1980), "Ecotoxicity in Canada."

**19 *The Energy We Eat***

The sleek brunette caught my eye in the April 30, 1979, issue of *Advertising Age*, which also contained the plastic cow now found in this chapter. Statistics of a different sort were obtained from *Synthetics and*

*Substitutes for Agricultural Products* (USDA, 1972) and *Food in Canada* (Sept. 1975). For more on the energy input of food, see *Technology Review* (Aug.-Sept. 1979), *Energy, Food and the Consumer* by Mary Rawitscher and Jean Mayer, and *Brandname Guide to Sugar,* by Ira Shannon.

**20 *The Other Economy***

The Stanford Research Institute Study referred to is *Voluntary Simplicity;* the most popular paper of its kind they've produced. The information on the Skills Exchange comes from first-hand experience—I am one of those teachers who substitutes human labor for physical resources. "Through the Organization Looking Glass" by Charles Handy (*Harvard Business Review,* Jan-Feb. 1980) gives more details on the trend to business decentralization.

**22 *The Freed Enterprise System***

I first came across figures of the startling production of collective farmers' plots when in the USSR in 1975 and confirmed them with a source at the West German Embassy in Moscow. The data is published in A. Yemelyanov's "The Agrarian Policy of the Party and Structural Advances in Agriculture," (*Problems in Economics,* March 1975, pp. 22-34), and reported in Hedrick Smith's, *The Russians.*

The Energy Probe study referred to in this chapter is Jack Gibbons' *The Cost of Not Implementing Marginal Cost Pricing.* The Mellon Institute study is entitled *The Least Cost Energy Strategy.*

**23 *The Unwritten Agenda***

Glaser, in *Physics Today* (July 1977, p. 66), argues that the use of solar satellites as weapons systems could be prevented "by making the satellite accessible to international inspection, or by arranging for international ownership." His conviction seems to wane, almost immediately, however, as he continues: "Furthermore, the vulnerability of the satellite is the most effective deterrent to its use as a weapon system. Once the benefit of satellite power systems can be provided on a worldwide basis, it will be in the common interest of all nations not to interfere with the operation of satellite power systems." The arguments have a familiar ring, in both peace and wartime, but a poor track record.

# INDEX